T0296704

CAMBRIDGE TRACTS IN MATHEMATICS

General Editors

B. BOLLOBAS, F. KIRWAN, P. SARNAK, C. T. C. WALL

94

The volume of convex bodies and Banach space geometry

GILLES PISIER

Université Paris VI

The volume of convex bodies and Banach space geometry

CAMBRIDGE UNIVERSITY PRESS
Cambridge, New York, Melbourne, Madrid, Cape Town, Singapore, São Paulo

Cambridge University Press
The Edinburgh Building, Cambridge CB2 8RU, UK

Published in the United States of America by Cambridge University Press, New York

www.cambridge.org
Information on this title: www.cambridge.org/9780521364652

First published 1989
First paperback edition 1999

A catalogue record for this publication is available from the British Library

Library of Congress Cataloguing in Publication data
Pisier, Gilles, 1950–
The volume of convex bodies and Banach space geometry
(Cambridge tracts in mathematics; 94)
Bibliography: p.
Includes indexes.
1. Banach spaces.   2. Inequalities (Mathematics)
I. Title. II. Series.
QA322.2.P57 1989   515.7'32   88-35356

ISBN 978-0-521-36465-2 hardback
ISBN 978-0-521-66635-0 paperback

Transferred to digital printing 2007

Contents

Introduction

The aim of these notes is to give a self-contained presentation of a number of recent results (mainly obtained during the period 1984–86) which relate the volume of convex bodies in $\mathbf{R}^n$ and the geometry of the corresponding finite-dimensional normed spaces.

The methods we employ combine traditional ideas in the Theory of Convex Sets (maximal volume ellipsoids, mixed volumes) together with Probability (Gaussian processes), Approximation Theory (Entropy numbers), and the Local Theory of Banach spaces (K-convexity, cotype and type).

During the last decade, considerable progress was achieved in the *Local Theory*, i.e. the part of Banach Space Theory which uses finite-dimensional (f.d. for short) tools to study infinite dimensional spaces.

The paper [FLM] was one of the first to relate the analytic properties of a Banach space (such as type and cotype) with the geometry of its f.d. subspaces. More precisely, let $B \subset \mathbf{R}^n$ be a ball (by this we mean a compact convex symmetric body admitting the origin as an interior point). In [FLM] the following question is studied:

Fix $\varepsilon > 0$, for which integer k does there exist a $(1 + \varepsilon)$-Euclidean section of B? Equivalently, for which k does there exist a subspace $F \subset \mathbf{R}^n$ with dimension k and an ellipsoid $D \subset F$ such that

$$(0.1) \qquad\qquad D \subset B \cap F \subset (1 + \varepsilon)D?$$

By a celebrated result of Dvoretzky, this is *always* true for $k = k(\varepsilon, n)$, with $k(\varepsilon, n) \to \infty$ when $n \to \infty$ (with $\varepsilon > 0$ fixed). More precisely ([M6]), this holds for $k = [\phi(\varepsilon) \operatorname{Log} n]$, with $\phi(\varepsilon) > 0$ depending only on ε. This estimate is the best possible in general; indeed, where B is the unit cube $B = [-1, 1]^n$ (i.e. the unit ball of ℓ_∞^n) then this Log n bound cannot be improved. One of the main discoveries of [FLM] is that if B and all its sections are far from cubes, in a certain analytic sense, then the estimate can be improved to

$$k(\varepsilon, n) = [\phi(\varepsilon)n^\alpha]$$

for some $\alpha > 0$.

Precisely, if the normed space E for which B is the unit ball has cotype $q(2 \leq q < \infty)$ with constant C then this holds with $\alpha = 2/q$ and $\phi(\varepsilon)$ depending only on ε and C.

The most striking case is the case $q = 2$ (hence $\alpha = 1$) for which we find in answer to (0.1) an integer k proportional to n. For instance, this case covers the case of $E = \ell_p^n$ for $1 \leq p < 2$. By duality, (0.1) implies that B°, the polar of the ball B, admits a projection onto F which is $(1 + \varepsilon)$-equivalent to an ellipsoid. Namely, if P_F is the orthogonal projection from $\mathbf{R}^n$ onto F, we have

$$(0.2) \qquad\qquad (1 + \varepsilon)^{-1} D^\circ \subset P_F(B^\circ) \subset D^\circ.$$

Since B is arbitrary, we can replace B by B° in (0.2). Thus, Dvoretzky's Theorem says that any n-dimensional body B admits both k-dimensional sections and k-dimensional projections which are almost ellipsoids with $k \to \infty$ when $n \to \infty$. However, in general k must remain small compared with n.

One of the striking recent discoveries of Milman [M1] is that if one considers the class of all projections of sections of B (instead of either projections or sections) then one can *always* find a projection of a section of B (say $P_{F_2}(F_1 \cap B)$ with $F_2 \subset F_1 \subset \mathbf{R}^n$) which is $(1 + \varepsilon)$-equivalent to an ellipsoid and has dimension $k = [\psi(\varepsilon)n]$ with $\psi(\varepsilon) > 0$ depending only on $\varepsilon > 0$.

Thus we find again k proportional to n, but this time without any assumption on B.

This surprising result gives the impression that in a number of questions an arbitrary ball in $\mathbf{R}^n$ should behave essentially like an ellipsoid. We will see several illustrations of this in the present volume. We now describe the contents.

This book has two distinct parts.

The objective of the first part (Chapters 1 to 8) is to present self-contained proofs of three fundamental results:

(I) *The quotient of subspace Theorem (Q.S.-Theorem for short) due to Milman [M1]: For each $0 < \delta < 1$ there is a constant $C = C(\delta)$ such that every n-dimensional normed space E admits a quotient of a subspace $F = E_1/E_2$ (with $E_2 \subset E_1 \subset E$) with dimension $\dim F \geq \delta n$ which is C-isomorphic to a Euclidean space.*

(II) *The inverse Santaló inequality due to Bourgain and Milman [BM]: There are positive constants α and β (independent of n) such that for all balls $B \subset \mathbf{R}^n$ we have*

$$\alpha/n \leq (\mathrm{vol}(B)\,\mathrm{vol}(B^\circ))^{1/n} \leq \beta/n.$$

(The upper bound goes back to a 1949 article by Santaló [Sa1].)

(III) *The inverse Brunn–Minkowski inequality due to Milman [M5]: Two balls B_1, B_2 in $\mathbf{R}^n$ can always be transformed (by a volume preserving linear isomorphism) into balls $\widetilde{B}_1, \widetilde{B}_2$ which satisfy*

$$\mathrm{vol}(\widetilde{B}_1 + \widetilde{B}_2)^{1/n} \le C \left[\mathrm{vol}(\widetilde{B}_1)^{1/n} + \mathrm{vol}(\widetilde{B}_2)^{1/n} \right]$$

where C is a numerical constant independent of n. Moreover, the polars $\widetilde{B}_1^\circ, \widetilde{B}_2^\circ$ and all their multiples also satisfy a similar inequality.

We present *two different approaches* to these results which can be read essentially independently. In Chapter 7, we reverse the chronological order; we prove (III) first and then deduce (II) and (I) as easy consequences.

In Chapter 8 we prove (I) by essentially the original method of [M1] and then give a very simple proof that (I) implies (II).

In the second part of the book (Chapters 10 to 15) we give a detailed exposition focused on the recently introduced classes of Banach spaces of weak cotype 2 or of weak type 2 and the intersection of these classes, the class of weak Hilbert spaces.

The previous chapters contain complete proofs of all the necessary ingredients for these results and various related estimates. Let us now review the contents of this book, chapter by chapter.

Chapter 1 introduces some terminology, notation, and preliminary background.

In Chapter 2 one of our main tools—the majorization of the Gaussian K-convexity constant of an n-dimensional space—is discussed in detail.

The connections between Gaussian measures and volume estimates are numerous. For instance, consider the following classical formula where B is any ball in $\mathbf{R}^n$, γ_n the canonical probability measure on $\mathbf{R}^n$ and D the Euclidean unit ball in $\mathbf{R}^n$:

$$(0.3) \quad \lim_{t \to \infty} \frac{\mathrm{vol}(B + tD) - \mathrm{vol}(tD)}{t^{n-1}\,\mathrm{vol}(D)} = c_n n^{1/2} \int \sup_{t \in B} <t, x> \, d\gamma_n(x),$$

where c_n is a constant tending to 1 when $n \to \infty$ (see Remark 1.5 and the beginning of Chapter 9 for more information). We will see more connections when we come to Chapter 5.

In Chapter 3 we present several properties of the maximal volume ellipsoids. This goes back to Fritz John who proved in a 1948 paper

[Joh] that any ball $B \subset \mathbf{R}^n$ contains a unique ellipsoid of maximal volume. As observed more recently by D. Lewis [L], one may impose on the class of ellipsoids D various different types of constraint (instead of $D \subset B$) and consider in each case the ellipsoid of maximal volume.

In particular, for a given ball B, we can consider the ellipsoid of maximal volume among all ellipsoids D such that the quantity (0.3) is ≤ 1. This ellipsoid (called the ℓ-ellipsoid in the sequel) plays a crucial role in the proofs of (I), (II), and (III) above.

In Chapter 4 we present the proof of Dvoretzky's Theorem and several related facts. We follow the usual *concentration of measure* approach (cf. [M6], [FLM]) but we use Gaussian measures instead of the Haar measure on the orthogonal group. This approach underlines the close connection between geometric characteristics of a space (here the dimension of the almost Euclidean sections of the unit ball) and Gaussian random variables.

In Chapter 5 we present results from the theory of Gaussian processes, mainly the Dudley–Sudakov Theorem (Theorems 5.6 and 5.5) which gives both a lower and an upper bound for integrals such as

$$(0.4) \qquad\qquad \mathbf{E} \sup_{t \in B} X_t$$

when (X_t) is a Gaussian process indexed by a set B. These estimates are given in terms of the metric $d(s,t) = \|X_t - X_s\|_2$ on B and involve the smallest number of balls of d-radius ε which are enough to cover B. Such bounds for (0.4) can be related to volume estimates, for instance via (0.3).

These inequalities can be reformulated in the language of entropy numbers of compact linear operators. We present a brief introduction to the theory of these numbers in Chapter 5. In particular, several results due to B. Carl will be very useful in subsequent chapters.

In Chapter 6 we present the notion of volume ratio. The volume ratio of an n-dimensional space E with unit ball B is defined as

$$\left(\frac{\mathrm{vol}(B)}{\mathrm{vol}(D_{\max})} \right)^{1/n},$$

where $D_{\max} \subset B$ is the maximal volume ellipsoid. This was introduced by Szarek [S], [ST] following earlier work of Kašin [Ka1] in Approximation Theory. Szarek observed that the ℓ_1^n-balls have a volume ratio uniformly bounded (when $n \to \infty$) and that this implies the striking orthogonal decomposition of ℓ_1^{2n} (due to Kašin) as $E_1 + E_2$ with E_1, E_2

both uniformly isomorphic to ℓ_2^n. This is now often called the Kašin decomposition of ℓ_1^n. We present Szarek's proof of this in Chapter 6 as well as several properties of spaces with bounded volume ratio.

To give the flavor of what goes on in this book, we wish to record here several interesting inequalities about the volume ratio. Let us denote by $\| \ \|_B$ the gauge of B. Let B_2 be the canonical Euclidean ball with its normalized surface measure σ on the boundary. Note that we have for any norm on $\mathbf{R}^n$

$$(0.5) \qquad \int \|x\|_B \, d\sigma(x) = c_n n^{-1/2} \int \|x\|_B \, d\gamma_n(x)$$

with c_n as above ($c_n \to 1$ when $n \to \infty$).

Integrating in polar coordinates, we have (see Chapter 6)

$$(0.6) \qquad \left(\frac{\mathrm{vol}(B)}{\mathrm{vol}(B_2)} \right)^{1/n} = \left(\int \|x\|_B^{-n} \, d\sigma(x) \right)^{1/n}.$$

By a classical inequality of Urysohn (see Chapter 1), we have

$$(0.7) \qquad \left(\int \|x\|_B^{-n} \, d\sigma(x) \right)^{1/n} \leq \int \|x\|_{B^\circ} \, d\sigma(x).$$

On the other hand, by convexity we have, obviously,

$$(0.8) \qquad \left(\int \|x\|_B \, d\sigma(x) \right)^{-1} \leq \left(\int \|x\|_B^{-n} \, d\sigma(x) \right)^{1/n}.$$

These inequalities (0.5) to (0.8) point at another close connection between volume ratio and Gaussian integrals. In particular, if we denote

$$\bigwedge = \int \|x\|_B \, d\sigma(x) \cdot \int \|x\|_{B^\circ} \, d\sigma(x),$$

then we find

$$\frac{1}{\bigwedge} \leq \left(\frac{\mathrm{vol}(B) \, \mathrm{vol}(B^\circ)}{\mathrm{vol}(B_2)^2} \right)^{1/n} \leq \bigwedge.$$

Now a simple computation shows that

$$\int \|x\|_B^2 \, d\sigma(x) = n^{-1} \int \|x\|_B^2 \, d\gamma_n(x).$$

Therefore

$$\bigwedge \leq n^{-1} \left(\int \|x\|_B^2 \, d\gamma_n(x) \cdot \int \|x\|_{B^\circ}^2 \, d\gamma_n(x) \right)^{1/2}.$$

But the product $\mathrm{vol}(B) \cdot \mathrm{vol}(B^\circ)$ is an "affine invariant", i.e. it does not change if we replace B by $u^{-1}(B)$ for any u in $GL(n)$ (we denote here by $GL(n)$ the set of all invertible linear transfomations on $\mathbf{R}^n$). Note that $\|x\|_{u^{-1}(B)} = \|ux\|_B$ and

$$\|x\|_{(u^{-1}B)^\circ} = \|u^{-1*}x\|_{B^\circ}.$$

Therefore, the preceding estimate can be rewritten as follows:

$$\frac{1}{\widetilde{\Lambda}} \leq \left(\frac{\mathrm{vol}(B)\,\mathrm{vol}(B^\circ)}{\mathrm{vol}(B_2)^2} \right)^{1/n} \leq \widetilde{\Lambda},$$

where

$$\widetilde{\Lambda} = \inf_{u \in GL(n)} n^{-1} \left(\int \|ux\|_B^2 d\gamma_n(x) \int \|u^{-1*}x\|_{B^\circ}^2 d\gamma_n(x) \right)^{1/2}.$$

Equivalently, with the notation of Chapter 3, if E is the normed space which admits B as its unit ball, we have

$$\widetilde{\Lambda} = \inf \left\{ n^{-1}\ell(u)\ell(u^{-1*}) \big| u : \ell_2^n \to E \quad u \text{ invertible} \right\}.$$

This shows that (II) (and similarly (I) or (III)) are easy to derive when $\widetilde{\Lambda}$ can be bounded above. This is possible in the K-convex case (see Chapter 3); namely, we prove in Chapter 3 that $\widetilde{\Lambda} \leq K(E)$ where $K(E)$ is the "K-convexity constant" of E, but unfortunately the constant $K(E)$ does not remain bounded when E and n are arbitrary. Nevertheless, the somewhat special properties of this constant $K(E)$ (as presented in Chapter 2) are among the most crucial tools in the proofs of (I), (II), or (III) above.

In Chapter 7 we present our proof of (III). The main ingredients are the results of Chapters 2, 3, 5, and 6.

In Chapter 8 we present Milman's proof of (I) and our proof of (II) as a simple consequence of (I). We also derive very directly from (II) the duality of entropy numbers for operators of rank n, due to König and Milman (see Theorem 8.10).

Both chapters can be read independently. In particular, the reader who wishes to read Chapter 8 before Chapter 7 should be warned that Chapter 7 is much simplified if one assumes (II) known. In particular, $M(B,D)$ reduces to just one of its factors if one already knows (II).

In Chapter 9 we study the volume numbers $v_n(T)$ of an operator $T : X \to \ell_2$. These numbers are defined as

$$v_n(T) = \sup \left(\frac{\mathrm{vol}(P(\overline{T(B_X)}))}{V_n} \right)^{1/n},$$

where the supremum runs over all orthogonal projections P of rank n on ℓ_2 and where we have denoted by V_n the volume of the canonical Euclidean unit ball.

We compare these numbers with the entropy numbers of T and show that they are almost equivalent. For instance, we show that if $\alpha < -1/2$ then

$$\limsup \frac{\text{Log } v_n(T)}{\text{Log } n} = \alpha \text{ iff } \limsup \frac{\text{Log } e_n(T)}{\text{Log } n} = \alpha.$$

This was conjectured by Dudley [Du1] and was established in [MP2]. The results of this chapter are not used in the sequel.

Chapters 10 to 15 constitute the second part of this book. They are motivated by the following property of certain Banach spaces X.

(*) *There is* $0 < \delta < 1$ *and a constant* C *such that every f.d. subspace* $E \subset X$ *contains a subspace* $F \subset E$ *with* $\dim F \geq \delta \dim E$ *which is* C-*isomorphic to a Euclidean space.*

As mentioned above, it was proved in [FLM] that this holds if X has cotype 2. Moreover, an example of Johnson (cf. [FLM]) shows that conversely (*) does not imply cotype 2, but only cotype $2 + \varepsilon$ for every $\varepsilon > 0$.

Independently of [FLM] and almost simultaneously, Kašin ([Ka1]) proved that (*) holds for $E = \ell_1^n$ (uniformly over n) and—following Szarek's proof [S]—this holds more generally if X has uniformly bounded volume ratio, i.e. if

$$(0.9) \qquad \sup\{vr(E) \mid E \subset X \ \dim E < \infty\} < \infty.$$

The question was then raised whether these two different approaches are related and specifically whether cotype 2 implies (0.9). This was proved by Bourgain and Milman in [BM]. Shortly after, this was continued (with somewhat different methods) by Milman and the author who showed in particular that (*) implies (0.9) (cf. [MP1]). In the same paper, a weakened version of the notion of cotype 2 (*weak cotype 2*) was introduced and was shown to be equivalent to (*) and (0.9).

Chapter 10 is mainly devoted to these results of [MP1].

In Chapter 11, we present the notion of *weak type 2* which is somewhat dual to the preceding. A space X is weak type 2 iff its dual X^* is K-convex and weak cotype 2, cf. [MP1]. We also include a characterization of weak type 2 in terms of volume ratio due to A. Pajor [Pa1].

In Chapter 12 we discuss the *weak Hilbert spaces*, i.e. the spaces which are both weak cotype 2 and weak type 2. This chapter is based on [P5]. For instance, we show that X is a weak Hilbert space iff one of the following properties (a), (b) holds.

(a) *There is a constant C such that for all f.d. subspaces $E \subset X$ there are ellipsoids D_1, D_2 in E such that*

$$D_1 \subset B_X \cap E \subset D_2$$

and

$$\left(\frac{\mathrm{vol}(D_2)}{\mathrm{vol}(D_1)}\right)^{1/n} \leq C.$$

(b) *There is a constant C such that for all n and all $x_1, \dots, x_n$, $x_1^*, \dots, x_n^*$ in the unit balls respectively of X and X^* we have*

$$\left|\det\left(<x_i^*, x_j>\right)\right|^{1/n} \leq C.$$

In Chapter 13 we present some examples of weak Hilbert spaces which are not isomorphic to Hilbert spaces. These examples are due to W. B. Johnson (see [FLM]) but are based on an earlier example of Tsirelson of a reflexive Banach space which contains ℓ_p for no p (cf. [T], [FJ2]). Since such spaces have an unconditional basis, it would be unreasonable to dismiss them as *pathological*. We refer the reader to the forthcoming book of Casazza and Shura [CS] for more information and similar examples.

In Chapter 14 we present some ideas due to W. B. Johnson which show that weak Hilbert spaces are reflexive, and hence super-reflexive.

In Chapter 15 we use the theory of determinants (sometimes called the Fredholm Theory) for elements of the projective tensor product $X^* \widehat{\otimes} X$ as described in [G2] to show that weak Hilbert spaces have the Approximation Property, and hence the Uniform Approximation Property. We also prove that the so-called Lidskii trace formula remains valid in a weak Hilbert space X: Let T be a nuclear operator on X; then, if its eigenvalues $\{\lambda_n\}$ are absolutely summable, we have

$$tr(T) = \sum \lambda_n.$$

In general, however, unless X is isomorphic to a Hilbert space, the eigenvalues of T are not absolutely summable (cf. [JKMR]).

If X is weak Hilbert, and if the eigenvalues are rearranged so that $|\lambda_1| \geq |\lambda_2| \geq, \ldots$, then any nuclear operator satisfies $\sup_n n|\lambda_n| < \infty$. In other words, instead of (λ_n) in ℓ_1 we find (λ_n) in *weak-ℓ_1* (the space $\ell_{1\infty}$) and this characterizes weak Hilbert spaces. Analogous statements hold for weak cotype 2 and weak type 2 (see [P5] for a broader notion of weak-P when P is a given property of Banach spaces). This explains in part the choice of the adjective *weak* common to all these notions.

We should warn the reader, however, that the space often called *weak-L_2* (often denoted $L_{2\infty}$) is not a weak Hilbert space; it is well known that this space contains an isomorphic copy of ℓ_∞.

In general, we do not give references in the text, but in the **Notes and Remarks** following each chapter. There are a number of exceptions to this rule, but, in any case, we warn the reader *always* to consult the **Notes and Remarks** section to find out to whom a given result should be credited.

We apologize in advance for possible errors or for references which may have been omitted for lack of accurate information.

This book is based on a series of lectures given at the University of Paris VI (spring 1985) and on graduate courses given in Paris VI (spring 1986) and at Texas A&M University (fall 1986). I am grateful to all those who participated in these lectures and especially to Alvaro Arias, who also did a careful proofreading of the first draft. I am also grateful to N. W. Naugle for his expert advice on TEX and for his help in dealing with diagrams. Finally, special thanks are due to Jan Want for her excellent typing.

This work was partly supported by the NSF.

Chapter 1

Notation and Preliminary Background

Let E, F be Banach spaces. If E, F are isomorphic, we define their *Banach–Mazur distance* as

$$(1.1) \qquad d(E, F) = \inf\{\|T\| \ \|T^{-1}\|\}$$

where the infimum runs all isomorphisms $T : E \to F$. If they are not isomorphic, we set $d(E, F) = +\infty$. If E, F are isomorphic and if $d(E, F) \le \lambda$, we will say briefly that E and F are λ-isomorphic (or that E is λ-isomorphic to F).

For all Banach spaces, E, F, G we have obviously

$$d(E, G) \le d(E, F) d(F, G).$$

We denote by B_E the unit ball of E and by I_E the identity operator on E.

In most of these notes we work with *finite dimensional* Banach spaces E, F. We will abreviate finite dimensional by f.d. The above *distance* is then particularly useful since any two spaces with the same dimension are isomorphic. Let $K(n)$ be the set of all normed (or Banach) spaces of dimension n. For E, F in $K(n)$, a simple compactness argument shows that the infimum is attained in (1.1) so that E, F are *isometric* iff $d(E, F) = 1$. The relation E *is isometric to* F is clearly an equivalence relation on $K(n)$. Let us denote by $\widetilde{K}(n)$ the set of all classes modulo this equivalence. Then it is not hard to check that $\widetilde{K}(n)$ equipped with the metric $\delta(E, F) = \mathrm{Log}\, d(E, F)$ is a compact metric space, sometimes called the Banach–Mazur compactum.

We will denote by ℓ_p^n the space $\mathbf{R}^n$ equipped with the norm $\|x\| = (\sum_1^n |x_i|^p)^{1/p}$. In particular, ℓ_2^n is the n-dimensional Euclidean space. The latter space plays *a central role* among the elements of $K(n)$ (or of $\widetilde{K}(n)$).

In particular, we show in Chapter 3 a classical result of F. John [Joh]

$$d(E, \ell_2^n) \le n^{1/2} \text{ for all } E \text{ in } K(n).$$

1

There has been in recent years a great deal of progress concerning the *local theory* of Banach spaces. This is the part of Banach space theory which uses mainly finite-dimensional tools and methods. In this theory an infinite-dimensional space is studied through the collection of all its finite-dimensional subspaces. For instance, by a fundamental theorem of Dvoretzky (cf. Chapter 4), every-infinite dimensional space X contains for each n and $\varepsilon > 0$ a subspace E such that $d(E, \ell_2^n) < 1 + \varepsilon$. Recently these methods have been successfully applied to prove several inequalities on the volume of convex symmetric bodies in $\mathbf{R}^n$. We will call these simply *balls*. More precisely, throughout the sequel, a *ball* will be a compact convex symmetric subset $B \subset \mathbf{R}^n$ with non-empty interior. Let $\| \ \|_B$ be the gauge of B. For any ball $B \subset \mathbf{R}^n$, the space $\mathbf{R}^n$ equipped with $\| \ \|_B$ is a Banach space admitting B as its unit ball. Conversely, any n-dimensional normed space E can be identified (in more than one way) with $\mathbf{R}^n$ and hence we may associate to E its unit ball $B_E \subset \mathbf{R}^n$. Note that two balls B_1, B_2 in $\mathbf{R}^n$ correspond to two isometric Banach spaces iff there is a linear isomorphism $u : \mathbf{R}^n \to \mathbf{R}^n$ such that $u(B_1) = B_2$.

If the normed space associated to a ball $B \subset \mathbf{R}^n$ is a Hilbert space, then B is an ellipsoid, i.e. there is an isomorphism $u : \mathbf{R}^n \to \mathbf{R}^n$ which maps B onto the canonical Euclidean ball $B_{\ell_2^n}$.

It will be useful to recognize geometrically the balls of subspaces and quotient spaces of an n-dimensional normed space E with unit ball $B_E \subset \mathbf{R}^n$. For subspaces this is immediate: let F be a subspace of E; clearly the *section* $F \cap B_E$ can be viewed as the unit ball of the normed subspace F. We may also consider the quotient space E/F. Geometrically, this space corresponds not to sections of B_E, but to linear projections of B_E. Indeed, let $P : E \to E$ be any linear projection such that $\ker P = F$. Let G be the range of P. We equip G with the norm which admits $P(B_E)$ as its unit ball. Then G is isometric to E/F. In particular, we may wish to use the orthogonal projection P_G into G, orthogonal with respect to a fixed scalar product on $\mathbf{R}^n$ (for instance the usual one). Then $P_G(B_E)$ can be naturally identified with the unit ball of the normed space $E/G^\perp$.

In the sequel, we always denote by E^* the dual space of E equipped with the dual norm. Recall that for $K \subset \mathbf{R}^n$, the polar set is defined as

$$K^\circ = \{x \in \mathbf{R}^n | \ <x, y> \ \le 1 \quad \forall \ y \in K\}.$$

Clearly, if $B_E \subset \mathbf{R}^n$ is the unit ball of E, we may identify B_{E^*} with the polar $(B_E)^\circ$ of B_E. We note the obvious identities for all subspaces $F \subset \mathbf{R}^n$:

$$(F \cap B_E)^\circ = P_F(B_E^\circ) \quad (P_F(B_E))^\circ = F \cap (B_E^\circ).$$

(Here the first and third polar sets are with respect to F with the induced scalar product.)

We now state and prove the Brunn–Minkowski inequality. Although we do not really need this inequality (or its corollaries) in these notes, it is natural to include it here.

Theorem 1.1. *Let A, B be two compact subsets of $\mathbf{R}^n$, then for all λ in [0,1] we have*

(1.1) $$\mathrm{vol}(\lambda A + (1 - \lambda)B) \geq \mathrm{vol}(A)^\lambda \, \mathrm{vol}(B)^{1-\lambda},$$

and

(BM) $$(\mathrm{vol}(A + B))^{1/n} \geq (\mathrm{vol}(A))^{1/n} + (\mathrm{vol}(B))^{1/n}.$$

We have learned the proof below from Keith Ball (cf. [Bal]). It is based on ideas from the papers [Pr] and [Le]. (cf. also [BrL]). We first prove a lemma.

Lemma 1.2. *Let f, g, ϕ be measurable functions from $\mathbf{R}^n$ into $[0, \infty]$ such that for some $0 < \lambda < 1$ and all r, s in $\mathbf{R}^n$ we have*

$$\phi(\lambda r + (1 - \lambda)s) \geq f(r)^\lambda g(s)^{1-\lambda}.$$

Then, denoting by m the Lebesgue measure on $\mathbf{R}^n$, we have

(1.2) $$\int \phi \, dm \geq \left(\int f \, dm \right)^\lambda \left(\int g \, dm \right)^{1-\lambda}.$$

Proof: We first consider the case $n = 1$. We may clearly assume f, g bounded and (by homogeneity) satisfying $\|f\|_\infty = \|g\|_\infty = 1$.

Note that for any $0 \leq a < 1$, we have obviously

$$\{\phi \geq a\} \supset \lambda\{f \geq a\} + (1 - \lambda)\{g \geq a\}$$

and since both sets on the right are non-empty, this implies (by the one-dimensional Brunn–Minkowski inequality!)

$$m\{\phi \geq a\} \geq \lambda\, m\{f \geq a\} + (1-\lambda)m\{g \geq a\}.$$

After integration in a over $[0, \infty]$, this implies

$$\int \phi\, dm = \int \{\phi \geq a\}\, dm \geq \lambda \int f\, dm + (1-\lambda) \int g\, dm;$$

hence (by the arithmetic–geometric mean inequality)

$$\geq \left(\int f\, dm\right)^{\lambda} \left(\int g\, dm\right)^{1-\lambda}.$$

This completes the proof of (1.2) for $n = 1$.

It is then easy to deduce the case of $\mathbf{R}^n$ by induction on n. Suppose $n > 1$ and assume the lemma proved for $n - 1$. Consider ϕ, f, g from $\mathbf{R}^n$ into $[0, \infty]$. Let $y \in \mathbf{R}$ be fixed, define $\phi_y : \mathbf{R}^{n-1} \to [0, \infty]$ by $\phi_y(t) = \phi(t, y)$ and similarly for f_y and g_y. Clearly, we have, if

$$y = \lambda y_1 + (1-\lambda)y_0 \quad (y_0, y_1 \in \mathbf{R}), \quad \phi_y(\lambda r + (1-\lambda)s) \geq f_{y_1}(r)^{\lambda} g_{y_0}(s)^{1-\lambda}$$

for any r, s in $\mathbf{R}^{n-1}$. Therefore by the induction hypothesis,

$$\int_{\mathbf{R}^{n-1}} \phi_y \geq \left(\int_{\mathbf{R}^{n-1}} f_{y_1}\right)^{\lambda} \left(\int_{\mathbf{R}^{n-1}} g_{y_0}\right)^{1-\lambda}.$$

Finally, applying (1.2) one more time (here for $n = 1$), we obtain

$$\int \phi\, dm = \int \left(\int_{\mathbf{R}^{n-1}} \phi_y\right) dy \geq \left(\int f\, dm\right)^{\lambda} \left(\int g\, dm\right)^{1-\lambda}.$$

This completes the proof of Lemma 1.2. ∎

Proof of Theorem 1.1: The inequality (1.1) follows immediately from (1.2) with

$$\phi = 1_{\lambda A + (1-\lambda)B}, \quad f = 1_A, \quad g = 1_B.$$

Then the Brunn–Minkowski inequality (BM) follows by setting

$$\lambda = (\mathrm{vol}(A))^{1/n}(\mathrm{vol}(A)^{1/n} + \mathrm{vol}(B)^{1/n})^{-1}.$$

Note that if $A' = \mathrm{vol}(A)^{-1/n}A$ and $B' = \mathrm{vol}(B)^{-1/n}B$, then (1.1) implies

$$(1.3) \qquad \mathrm{vol}(\lambda A' + (1-\lambda)B') \geq 1,$$

and, since

$$\lambda A' + (1-\lambda)B' = \frac{A+B}{(\mathrm{vol}(A)^{1/n} + \mathrm{vol}(B)^{1/n})},$$

(1.3) immediately implies (BM) by homogeneity. ∎

Remark: For a different proof of (BM) and more information, we refer the interested reader to [Be], [BZ], [Eg] [BF], [Ha1] and [Ha2]. The reader who wishes to learn about recent developments in the Classical Theory of Convex Sets should consult for instance [Sa2] and the collection of surveys [GW].

As a corollary, we can prove the classical isoperimetric inequality in $\mathbf{R}^n$. We recall that the area of the boundary of a convex compact subset of $\mathbf{R}^n$ can be defined by an approximating procedure from the simpler case of polytopes (cf. e.g. [Be] or [Eg].)

Corollary 1.3. (Isoperimetric inequality) *Let C be a compact convex subset of $\mathbf{R}^n$ with non-empty interior. Let us denote by $a(C)$ the area of its boundary. Let B_2 be the canonical euclidean ball in $\mathbf{R}^n$. Then*

$$\left(\frac{\mathrm{vol}(C)}{\mathrm{vol}(B_2)}\right)^{1/n} \leq \left(\frac{a(C)}{a(B_2)}\right)^{1/n-1}.$$

Proof: Indeed, the area can be *derived* from the volume in a simple way, since we have

$$(1.4) \qquad a(C) = \lim_{t \to 0} t^{-1}(\mathrm{vol}(C + tB_2) - \mathrm{vol}(C)).$$

By (1.4), we have $a(B_2) = n \ \mathrm{vol}(B_2)$ and by (BM)

$$\mathrm{vol}(C + tB_2) \geq (\mathrm{vol}(C)^{1/n} + t \ \mathrm{vol}(B_2)^{1/n})^n$$

$$\geq \mathrm{vol}(C) + nt \ \mathrm{vol}(B_2)^{1/n} \ \mathrm{vol}(C)^{\frac{n-1}{n}} + o(t) \ ;$$

hence

$$a(C) \geq \left(\frac{\mathrm{vol}(C)}{\mathrm{vol}(B_2)}\right)^{\frac{n-1}{n}} \cdot a(B_2). \quad ∎$$

We can also deduce from (BM) a classical inequality of Urysohn [U] which we do not really use in the sequel, but which clarifies the relationship between our methods and classical inequalities such as (BM). We denote by S the Euclidean unit sphere of $\mathbf{R}^n$ and by σ its normalized area measure.

Corollary 1.4. *(Urysohn's Inequality) Let K be a compact subset of $\mathbf{R}^n$. Let*

$$\|x\|_{K^\circ} = \sup\{<x,y>, y \in K\}$$

for all $x \in \mathbf{R}^n$. Then

$$\left(\frac{\text{vol}(K)}{\text{vol}(B_2)}\right)^{1/n} \le \int_S \|x\|_{K^\circ}\, d\sigma(x).$$

Proof: It is easy to generalize (BM) to a finite number of sets $A_1, \dots, A_k$ so that for any numbers $m_i \ge 0$ we have

$$\sum m_i \text{vol}(A_i)^{1/n} \le \left(\text{vol}\left(\sum m_i A_i\right)\right)^{1/n}.$$

More generally (under suitable assumptions), if A_t depends on a parameter t varying in a measure space $(\Omega, \sum, \nu)$ then we have

$$(1.5) \qquad \int \text{vol}(A_t)^{1/n}\, d\nu(t) \le \left(\text{vol}\left(\int A_t\, d\nu(t)\right)\right)^{1/n}.$$

Now let (Ω, ν) be the orthogonal group $O(n)$ equipped with its normalized Haar measure. Let $A_t = t(K)$ for all t in $O(n)$. Then $\text{vol}(A_t) = \text{vol}(K)$ for all t. On the other hand, $\int t(K)\, d\nu(t)$ is clearly (by symmetry) a multiple of the Euclidean ball B_2. Hence

$$(1.6) \qquad \int t(K)\, d\nu(t) = \lambda B_2$$

for some number $\lambda \ge 0$.

Now let ξ be a fixed element of S. (For instance, let $\xi = e_1$ the first basis vector of $\mathbf{R}^n$.) Writing that the images under ξ of both sides of (1.6) coincide, we find

$$\int \xi(t(K))\, d\nu(t) = \lambda[-1, 1].$$

Clearly, $\xi(t(K)) = [a_t, b_t]$ with $a_t = \inf_{x \in K} \xi(t(x))$ and $b_t = \sup_{x \in K} \xi(t(x))$. This implies

$$\int b_t\, d\nu(t) = \lambda,$$

so that (1.5) implies

$$\text{vol}(K)^{1/n} \le \lambda(\text{vol}(B_2))^{1/n}.$$

This completes the proof since $\lambda = \int \|t^* \xi\|_{K^\circ} d\nu(t) = \int \|x\|_{K^\circ} d\sigma(x)$. ∎

The preceding proof was shown to me by V. Milman.

Remark 1.5: Let γ be the canonical Gaussian probability measure on $\mathbf{R}^n$. Then, integrating in polar coordinates, we find

$$\int \|x\|_{K^\circ} \, d\gamma(x) = c_n \int_S \|x\|_{K^\circ} \, d\sigma(x),$$

with $c_n = \int (\sum_1^n x_i^2)^{1/2} \, d\gamma(x) \le \sqrt{n}$ and $\frac{c_n}{\sqrt{n}} \to 1$ when $n \to \infty$. Thus, Corollary 1.5 implies *a fortiori*

$$\left(\frac{\text{vol}(K)}{\text{vol}(B_2)} \right)^{1/n} \le n^{-1/2} \left(\int \|x\|_{K^\circ}^2 \, d\gamma(x) \right)^{1/2}.$$

This last inequality explains typically why the estimates of Gaussian integrals of the form $\int \|x\|^2 \, d\gamma(x)$ (for some norm $\| \ \|$ on $\mathbf{R}^n$) can be used to evaluate certain volumes. In the present notes, we will exploit the recent techniques of the local theory of Banach spaces concerning integrals such as $\int \|x\|^2 d\gamma(x)$ to obtain some new inequalities on the volumes of convex bodies in $\mathbf{R}^n$.

In the second part of this chapter, we recall some basic facts of operator theory. An operator $T : H_1 \to H_2$ between Hilbert spaces is called a Hilbert–Schmidt operator if for some (equivalently for all) orthonormal basis (e_i) of H_1 we have $\sum \|T \, e_i\|^2 < \infty$. The *Hilbert–Schmidt norm* of T is then defined as

$$\|T\|_{HS} = \left(\sum \|T \, e_i\|^2 \right)^{1/2}.$$

Let (f_j) be an orthonormal basis in H_2. We have

$$(1.7) \qquad \|T\|_{HS}^2 = \sum_{ij} \left| < Te_i, f_j > \right|^2 = \sum_j \|T^* f_j\|^2 = \|T^*\|_{HS}^2.$$

This shows that $\|T\|_{HS}$ is independent of the choice of the basis (e_i) and we have

$$(1.8) \qquad \qquad \|T\|_{HS} = \|T^*\|_{HS}.$$

We also note the identity

$$(1.9) \qquad \begin{aligned} \|T\|_{HS}^2 &= \sum < Te_i, Te_i >= \sum < T^* Te_i, e_i > \\ &= \text{tr } T^* T = \text{tr } |T|^2 \end{aligned}$$

which shows that if T is Hilbert–Schmidt, then $T^*T = |T|^2$ is a trace class operator.

Recall that a Hilbert–Schmidt operator is a *fortiori* compact. For any compact operator T, we will denote by $(\lambda_i(T))_{i\geq 1}$ the eigenvalues of T repeated according to their multiplicities and arranged so that the sequence $\{|\lambda_n(T)|, n \geq 0\}$ is non-increasing. Let $T = U|T|$ be the polar decomposition of T where $|T| = (T^*T)^{1/2}$ is a positive self-adjoint operator and U is a partial isometry (it is isometric on the range of $|T|$).

Then clearly $\|T\|_{HS} < \infty$ iff $\||T|\|_{HS} < \infty$, and we have

$$(1.10) \qquad \|T\|_{HS} = \||T|\|_{HS}.$$

From (1.9) we derive

$$(1.11) \qquad \|T\|_{HS} = \left(\sum_1^\infty \lambda_n(|T|)^2\right)^{1/2}.$$

The numbers $\lambda_n(|T|)$ can also be characterized as the approximation numbers of T. Let $T : X \to Y$ be an operator between Banach spaces. We denote as usual

$$a_n(T) = \inf\{\|T - S\| \mid S : X \to Y \ rk(S) < n\}.$$

Note that $\|T\| = a_1(T) \geq a_2(T) \geq \dots$. In the particular case of an operator $T : H_1 \to H_2$ between Hilbert spaces, it is well known that

$$(1.12) \qquad a_n(T) = \lambda_n(|T|),$$

so that

$$(1.13) \qquad \|T\|_{HS} = \left(\sum a_n(T)^2\right)^{1/2}.$$

In the Banach space case, the class of *p-summing operators* introduced by Pietsch following the work of Grothendieck (cf. [Pi1]) replaces quite often (especially when $p = 2$) the class of Hilbert–Schmidt operators. Let $0 < p < \infty$. Recall that an operator $T : X \to Y$ between Banach spaces is called p-summing if there is a constant C such that for all finite sequences (x_i) in X, we have

$$\left(\sum \|Tx_i\|^p\right)^{1/p} \leq C \ \sup\left\{\left(\sum |\xi(x_i)|^p\right)^{1/p} \Big| \xi \in B_{X^*}\right\}.$$

We denote by $\pi_p(T)$ the smallest constant C for which this holds and by $\Pi_p(X,Y)$ the set of all p-summing operators from X into Y. This set is a Banach space when equipped with the norm π_p. These operators form an *operator ideal* in the sense of Pietsch, which essentially means that for all Banach spaces X_1, Y_1 and all operators $u : X_1 \to X$ and $v : Y \to Y_1$, if $T : X \to Y$ is p-summing, then vTu is also and we have

(1.14) $$\pi_p(vTu) \le \|v\|\pi_p(T)\|u\|.$$

This is easy to check from the definition. Moreover, if $T : X \to Y$ is p-summing and if ϕ is in $L_p(\Omega, \mu; X)$ (here (Ω, μ) denotes an arbitrary measure space) we have

(1.15) $$\|T(\phi)\|_{L_p(Y)} \le \pi_p(T) \sup\{\|\xi(\phi)\|_p | \xi \in B_{X^*}\}.$$

Indeed, this is easy to check for step functions and it can be extended to $L_p(X)$ by density. The space $L_p(\Omega, \mu; X)$ always denotes in the sequel the completion of $L_p(\Omega, \mu) \otimes X$ with the usual norm.

We will use the following well-known fact:

Proposition 1.6. *Let* $T : H_1 \to H_2$ *be an operator between two Hilbert spaces. Then* T *is 2-summing iff* T *is Hilbert–Schmidt and we have*
$$\pi_2(T) = \|T\|_{HS}.$$

Proof: We will prove the *only if* part first. So consider a 2-summing operator $T : H_1 \to H_2$ and let (e_i) be an orthonormal basis of H_1. We have

$$\left(\sum \|T\,e_i\|^2\right)^{1/2} \le \pi_2(T) \sup\left\{\left(\sum |\xi(e_i)|^2\right)^{1/2} \Big| \|\xi\| \le 1\right\}$$

$$\le \pi_2(T).$$

Hence, T is Hilbert–Schmidt and $\|T\|_{HS} \le \pi_2(T)$.

Conversely, assume that T is Hilbert–Schmidt. Let $x_1, \dots, x_n$ be arbitrary in H_1 and let (f_j) be an orthonormal basis of H_2. We have

$$\sum \|Tx_i\|^2 = \sum_{ij} |<f_j, Tx_i>|^2$$

$$= \sum_j \left(\sum_i |<T^*f_j, x_i>|^2\right)$$

$$\le \sum_j \|T^*f_j\|^2 \sup\left\{\sum |\xi(x_i)|^2 \Big| \|\xi\| \le 1\right\}.$$

This shows that T is 2-summing and by (1.7) we have $\pi_2(T) \leq \|T\|_{HS}$. ∎

As a corollary, we have

Corollary 1.7. *Let E_1, E_2 be Banach spaces which are isomophic respectively to the Hilbert spaces H_1 and H_2. Then every 2-summing operator $T : E_1 \to E_2$ satisfies the expression*

$$(1.16) \qquad \left(\sum a_n(T)^2 \right)^{1/2} \leq d(E_1, H_1) d(E_2, H_2) \pi_2(T).$$

Proof: This is immediate using (1.14) and (1.13) together with Proposition 1.6.

For various purposes, we will also need the following simple fact first used in a similar context by D. Lewis.

Lemma 1.8. *Let $T : H \to X$ be an operator from a Hilbert space into a Banach space. For every $\varepsilon > 0$, there is an orthonormal system (f_n) in H such that $\|T f_n\| \geq a_n(T) - \varepsilon$ for all $n \geq 1$. Moreover, if H is f.d., we have this also for $\varepsilon = 0$.*

Proof: Let $f_1 \in H$ be such that $\|f_1\| = 1$ and

$$\|T(f_1)\| \geq \|T\| - \varepsilon = a_1(T) - \varepsilon.$$

We consider the restriction of T to $S_1 = [f_1]^{\perp}$ and note that obviously $\|T_{|S_1}\| \geq a_2(T)$. Therefore, there is a norm 1 element $f_2 \in [f_1]^{\perp}$ such that

$$\|T(f_2)\| \geq a_2(T) - \varepsilon.$$

We then restrict T to $[f_1, f_2]^{\perp}$. Continuing in this way, we obtain a sequence (f_n) which has the announced property. In the f.d. case, since the unit balls are compact, we may take $\varepsilon = 0$. ∎

With this result we can improve Corollary 1.7 as follows.

Corollary 1.9. *Let $u : H \to X$ be a 2-summing operator from a Hilbert space into a Banach space. Then $(\sum a_k(u)^2)^{1/2} \leq \pi_2(u)$.*

Proof: Let $\varepsilon > 0$. Let (f_k) be an orthonormal basis such that $\|u f_k\| \geq a_k(u) - \varepsilon$. Then

$$\left(\sum \|u f_k\|^2 \right)^{1/2} \leq \pi_2(u) \sup \left\{ \left(\sum | < \xi, f_k > |^2 \right)^{1/2} \|\xi\| \leq 1 \right\}.$$

This immediately implies Corollary 1.9. ∎

Since we will be constantly estimating the volumes of certain balls in $\mathbf{R}^n$, it might be worth while to indicate how to compute the volumes of the unit balls of the most classical spaces such as ℓ_p^n:

For $x \in \mathbf{R}^n$, let $\|x\|_p = (\sum_1^n |x_i|^p)^{1/p}$. Let us denote

$$V_n(p) = \text{vol}\left(B_{\ell_p^n}\right)$$
$$= \text{vol}\{x \in \mathbf{R}^n \| x\|_p \leq 1\}.$$

This number is easy to compute using the following integral:

$$I_p = \int_{\mathbf{R}^n} \exp(-\|x\|_p^p) dx.$$

Indeed, we clearly have on one hand

$$I_p = \left(\int_{\mathbf{R}} \exp(-|t|^p) dt\right)^n$$

and on the other hand

$$I_p = \int \left(\int_{\|x\|_p}^\infty \frac{d}{dt}\left(-\exp(-t^p)\right) dt\right) dx$$
$$= \int_0^\infty pt^{p-1} \exp(-t^p) \ \text{vol}\{x|\|x\|_p < t\} dt$$
$$= V_n(p) \int_0^\infty pt^{n+p-1} \exp(-t^p) dt.$$

Therefore, we conclude that

$$V_n(p) = \frac{\left(2 \int_o^\infty \exp(-t^p) \ dt\right)^n}{\int_0^\infty pt^{n+p-1} \exp(-t^p) dt}.$$

After an elementary computation we can rewrite this using the gamma function (recall $\Gamma(\alpha) = \int_0^\infty t^{\alpha-1} \exp(-t) \ dt$) as follows:

(1.17) $$V_n(p) = \frac{\left(2\Gamma\left(1 + \frac{1}{p}\right)\right)^n}{\Gamma\left(1 + \frac{n}{p}\right)}.$$

In particular, this implies that there are positive constants $c_1(p)$ and $c_2(p)$ such that for all integers $n \geq 1$ we have $(0 < p < \infty)$:

(1.18) $$c_1(p)n^{-1/p} \leq \left(\text{vol}\left(B_{\ell_p^n}\right)\right)^{1/n} \leq c_2(p)n^{-1/p}.$$

The volume of the balls of the non-commutative finite-dimensional L_p-spaces is computed in [StR2].

Notes and Remarks

Most of the references are given in the text. We refer to [LT1] and [LT2] for more background on the geometry of Banach spaces, and to [Pi1] and [Kö] for more on operator theory and approximation numbers.

Chapter 2

Gaussian Variables. K-Convexity

Throughout this book we will denote by $(\Omega, \mathcal{A}, \mathbf{P})$ a suitable probability space. We will consider a sequence $(g_n)_{n \geq 1}$ of independent identically distributed (i.i.d. for short) Gaussian random variables (r.v. for short) with the standard normal distribution. This means that $\{g_n\}$ are independent and for each n we have

$$\forall \, t \in \mathbf{R} \qquad \mathbf{P}(g_n > t) = (2\pi)^{-1/2} \int_t^\infty e^{-x^2/2} \, dx.$$

More generally, for any Borel subset $B \subset \mathbf{R}^n$ we have

$$\mathbf{P}(\{(g_1, \ldots, g_n) \in B\}) = (2\pi)^{-n/2} \int_B \exp\left(-\frac{1}{2} \sum_1^n x_i^2\right) dx_1 \ldots dx_n.$$

The property which will be of crucial importance in the sequel is the rotational invariance of the Gaussian distribution.

Namely, if (a_{ij}) is an orthogonal $n \times n$ matrix, then the sequence

$$\tilde{g}_i = \sum_{j=1}^n a_{ij} g_j \qquad i = 1, 2, \ldots, n$$

has the same distribution on $\mathbf{R}^n$ as the original sequence $(g_1, \ldots, g_n)$.

This means that

(2.1) $$\mathbf{P}((\tilde{g}_1, \ldots, \tilde{g}_n) \in B) = \mathbf{P}\,(g_1, \ldots, g_n) \in B)$$

for any Borel subset B of $\mathbf{R}^n$.

This basic fact (2.1) is an immediate consequence of the invariance under $O(n)$ of the standard Gaussian probability measure on $\mathbf{R}^n$, which we denote by γ_n.

(2.2) $$\gamma_n = \exp\left(-\frac{1}{2} \sum_{i=1}^n x_i^2\right) dx_1 \ldots dx_n.$$

13

Let X be a random variable with values in a separable Banach space E. We will denote by $\mathrm{dist}(X)$ the distribution of X, i.e. the probability measure on E defined, for all Borel subset $B \subset E$, by $\mathrm{dist}(X)(B) = \mathbf{P}(X \in B)$.

Recall that if X and Y have the same distribution, then we have $\mathbf{E}\varphi(X) = \mathbf{E}\varphi(Y)$ for all Borel measurable functions $\varphi : E \to \mathbf{R}$ such that $\varphi(X)$ is integrable.

The identity (2.1) has many consequences. For instance, for any $(\alpha_1, \ldots, \alpha_n)$ in $\mathbf{R}^n$, we have

$$(2.3) \qquad \mathrm{dist}(\alpha_1 g_1 + \ldots + \alpha_n g_n) = \mathrm{dist}\left(g_1 \cdot \left(\sum_1^n |\alpha_i|^2\right)^{1/2}\right).$$

Indeed, we may, by homogeneity, assume $\sum \alpha_i^2 = 1$ and then we just use the fact that there is an A in $O(n)$ which maps $\alpha = (\alpha_1, \ldots, \alpha_n)$ into e_1 (the first vector of the canonical basis of $\mathbf{R}^n$).

This formula (2.3) can also be deduced from the classical Fourier transform formula

$$(2.4) \qquad \forall \xi \in \mathbf{R} \quad \mathbf{E}e^{i\xi g_j} = \int e^{i\xi x - x^2/2}\, dx (2\pi)^{-1/2} = e^{-\xi^2/2},$$

for all $j \geq 1$.

By (2.3) (and the remark preceding it) we have

$$(2.5) \qquad \left\|\sum_1^n \alpha_i g_i\right\|_p = \|g_1\|_p \left(\sum_1^n |\alpha_i|^2\right)^{1/2}$$

for all $0 < p < \infty$.

Let $\gamma(p) = \|g_1\|_p$. Clearly $\|g_1\|_p < \infty$ for all $p < \infty$. Moreover, we have

$$(2.6) \qquad \gamma(p) \leq K\, p^{1/2} \quad \text{if } 1 \leq p < \infty,$$

for some numerical constant K.

By (2.5), the closed span of $\{g_n\}$ in $L_p(\Omega, \mathcal{A}, \mathbf{P})$ is isometric to ℓ_2. Indeed, if (e_n) is the canonical basis of ℓ_2, we can define an isometric embedding $J : \ell_2 \to L_p$ which maps e_n into g_n for all n. By (2.5), the range of J (i.e. $J(\ell_2)$) is independent of $0 < p < \infty$. Let G be the closed span of $\{g_n\}$ in L_2. Then, by (2.5), $G = J(\ell_2)$, G is closed in all L_p spaces for all $0 < p < \infty$, and we have

$$(2.7) \qquad \forall\, x \in G \qquad \|x\|_p = \gamma(p)\|x\|_2.$$

It will be of interest to note that the closed subspace $G \subset L_p$ is *complemented* in L_p for $1 < p < \infty$.

Let $Q_1 : L_2 \to G$ be the orthogonal projection. Then Q_1 is bounded also from L_p onto G for $1 < p < \infty$. Indeed, if $2 \le p < \infty$, for any x in L_2, we have by (2.7)

$$\gamma(p)^{-1} \|Q_1 x\|_p = \|Q_1 x\|_2 \le \|x\|_2 \le \|x\|_p;$$

therefore $Q_1 : L_p \to L_p$ is bounded and $\|Q_1\|_{L_p \to L_p} \le \gamma(p)$. For $1 < p \le 2$, using duality and the self-adjointness of Q_1, we find

$$\|Q_1\|_{L_p \to L_p} \le \gamma(p').$$

The projection Q_1 plays a very important role in the theory that we will present here. (Equivalently, one can consider the orthogonal projection onto the span of the Rademacher functions; see below.) This projection enjoys a special property which will be crucial in the sequel. One way to formulate this property is as follows. We denote by $\mathbf{N}_*$ the set of all *non-zero* integers. We will say that $(\Omega, \mathcal{A}, \mathbf{P})$ is standard if $\Omega = \mathbf{R}^{\mathbf{N}_*}$, $\mathcal{A}$ is the Borel σ–algebra, and $\mathbf{P}$ is the canonical Gaussian probability γ_∞ on $\mathbf{R}^{\mathbf{N}_*}$.

Note then that the coordinate functions $x \to x_n$ (from $\mathbf{R}^{\mathbf{N}_*}$ into $\mathbf{R}$) form an i.i.d. sequence of normal Gaussian r.v.'s.

Lemma 2.1. *Assume* $(\Omega, \mathcal{A}, \mathbf{P})$ *standard. Then there is a sequence of orthogonal projections* $\{Q_k | k \ge 0\}$ *on* $L_2(\Omega, \mathbf{P})$ *(denoted simply by* L_2*) such that*

$$I = \sum_{k \ge 0} Q_k,$$

$$Q_0(\zeta) = \int \zeta \, d\mathbf{P} \quad \text{for all } \zeta \text{ in } L_2,$$

and such that for all ε *in* $[-1, 1]$ *the operator*

$$T(\varepsilon) = \sum_{k \ge 0} \varepsilon^k Q_k$$

is a positive contraction on L_2.

For Q_k we can take the orthogonal projection onto the span of the Hermite polynomials of degree exactly k on $\mathbf{R}^{\mathbf{N}_*}$. To check Lemma

2.1, we need to recall first some properties of Hermite polynomials. We refer e.g. to [N] for more details. The Hermite polynomials in one real variable x will be denoted by $(h_n(x))_{n \geq 0}$. They are determined by the generating formula

$$(2.8) \qquad \forall \, \lambda \in \mathbf{R} \quad \exp\left(\lambda x - \frac{\lambda^2}{2}\right) = \sum_{n \geq 0} \frac{\lambda^n}{n!} h_n(x).$$

Note that $h_o(x) \equiv 1$ and $h_1(x) = x$.

The sequence $(h_n)_{n \geq 0}$ is an orthogonal basis of $L_2(\mathbf{R}, \gamma_1)$. ($\gamma_1$ is defined in (2.2) above.)

More generally, the collection

$$\{h_{\alpha_1}(x_1) \cdot h_{\alpha_2}(x_2) \ldots \cdot h_{\alpha_n}(x_n) | (\alpha_1, \ldots, \alpha_n) \in \mathbf{N}^n\}$$

forms an orthogonal basis of $L_2(\mathbf{R}^n, \gamma_n)$. We need more notation to extend this to $\mathbf{R}^{\mathbf{N}*}$.

We will denote simply by A the set of all $\alpha = (\alpha_1, \alpha_2, \ldots)$, with $\alpha_n \in \mathbf{N}$ such that only finitely many α_n's are non zero.

Let $|\alpha| = \sum_{i \geq 1} |\alpha_i|$ and $\alpha! = \alpha_1! \alpha_2! \ldots$. As usual, we denote by $\mathbf{R}^{(\mathbf{N}*)}$ the space of all sequences $(\lambda_n)_{n \geq 1}$ with $\lambda_n \in \mathbf{R}$ such that only finitely many λ_n's are non zero. Then, for x in $\mathbf{R}^{\mathbf{N}*}$ we write

$$< \lambda, x > = \sum_{n=1}^{\infty} \lambda_n x_n,$$

and, for α in A,

$$\lambda^\alpha = \lambda_1^{\alpha_1} \lambda_2^{\alpha_2} \ldots .$$

Now, we can define, for all α in A,

$$\forall \, x \in \mathbf{R}_*^{\mathbf{N}} \quad H_\alpha(x) = h_{\alpha_1}(x_1) h_{\alpha_2}(x_2) \ldots ,$$

and we have a multinomial generating formula generalizing (2.8):
$$(2.9)$$
$$\forall \, x \in \mathbf{R}^{\mathbf{N}*} \quad \forall \, \lambda \in \mathbf{R}^{\mathbf{N}*} \quad \exp\left(< \lambda, x > - \frac{1}{2} \sum \lambda_n^2\right) = \sum_{\alpha \in A} \frac{\lambda^\alpha}{\alpha!} H_\alpha(x).$$

One then checks easily that $\{H_\alpha | \alpha \in A\}$ is an orthogonal basis of $L_2(\mathbf{R}^{\mathbf{N}}, \gamma_\infty)$ (i.e. of $L_2(\Omega, \mathbb{P})$ with $(\Omega, \mathbf{P})$ standard).

Let H_k be the closed span in L_2 of the functions $\{H_\alpha | \, |\alpha| = k\}$. This is called the *Wiener chaos* of degree k. Let $Q_k : L_2 \to H_k$ be the orthogonal projection. Clearly, $Q_0(\zeta) = \int \zeta d\mathbf{P}$ for all ζ in L_2. Moreover, when $k = 1$ (recall $h_1(x_n) = x_n$), H_1 is the span in L_2 of the sequence $\{x_n | n \geq 1\}$ and Q_1 (as above) the projection onto the span of the Gaussian i.i.d. sequence $\{x_n | n \geq 1\}$ on $(\mathbf{R^{N_*}}, \gamma_\infty)$.

It will be convenient to use the fact that the collection of functions

$$\begin{cases} \varphi_\lambda(x) = \exp\left(<\lambda, x> -\frac{1}{2}\sum_n \lambda_n^2\right) \\ \lambda \in \mathbf{R^{(N_*)}} \end{cases}$$

is total in L_2, so that if two linear operators coincide on $\{\varphi_\lambda | \lambda \in \mathbf{R^{(N_*)}}\}$ then they coincide on L_2. (The density of the span of $\{\varphi_\lambda\}$ follows from the Stone-Weierstrass Theorem.)

Proof of Lemma 2.1: Let S be the linear span of $\{H_\alpha | \alpha \in A\}$. Equivalently, S is the space of all polynomials in the variables x_1, $x_2, \ldots$. We will write simply L_p for $L_p(\Omega, \mathbf{P})$. Note that S is dense in L_p for all $0 < p < \infty$.

Let $-1 \leq \varepsilon \leq 1$. We introduce the operator $T(\varepsilon)$ defined on S by the formula

$$(2.10) \qquad \forall f \in S \quad T(\varepsilon)f(x) = \int f(\varepsilon x + (1 - \varepsilon^2)^{1/2}y)\mathbf{P}(dy).$$

Clearly $T(\varepsilon)$ transforms polynomials into polynomials and is linear. Moreover, for any $1 \leq p < \infty$ we have

$$\|T(\varepsilon)f\|_{L_p} \leq \left(\int\int |f(\varepsilon x + (1 - \varepsilon^2)^{\frac{1}{2}}y)|^p \, d\mathbf{P}(x) \, d\mathbf{P}(y)\right)^{1/p} = \|f\|_{L_p}.$$

The last equality follows from (2.3) (note that the random variable $(x, y) \rightarrow \varepsilon x + (1 - \varepsilon^2)^{1/2}y$ defined on $\mathbf{R^{N_*}} \times \mathbf{R^{N_*}}$ equipped with $\mathbf{P} \times \mathbf{P}$ admits $\mathbf{P}$ as its distribution).

This shows that (2.10) makes sense for f in L_p for almost every x and $T(\varepsilon)$ extends to a linear contraction on L_p for $1 \leq p < \infty$. Note that $T(\varepsilon)$ is positive in the sense that $T(\varepsilon)f \geq 0$ for all $f \geq 0$.

It remains to check that

$$T(\varepsilon) = \sum_{k \geq 0} \varepsilon^k Q_k.$$

For that purpose, consider φ_λ as above.

It is very easy to check that

$$T(\varepsilon)\varphi_\lambda = \varphi_{\varepsilon\lambda}.$$

On the other hand, by (2.9), we have

$$\left(\sum_{k\geq 0}\varepsilon^k Q_k\right)\varphi_\lambda = \sum_{\alpha\in A}\varepsilon^{|\alpha|}\frac{\lambda^\alpha}{\alpha!}H_\alpha;$$

hence by (2.9) again

$$= \varphi_{\varepsilon\lambda}.$$

This shows that $T(\varepsilon)$ coincides with $\sum \varepsilon^k Q_k$ on $\{\varphi_\lambda\}$; hence it coincides with it on L_2. ∎

We now come to the notion of K-convexity which was introduced in [MaP] and further studied in [P3]. In order to introduce this notion, we need to clarify a notation. Let (Ω, m) be any measure space, let $L_2 = L_2(\Omega, m)$, let T be an operator on L_2 and let X be a Banach space. We denote by I_X the identity operator on X. Then clearly $T \otimes I_X$ is a linear operator well defined on $L_2 \otimes X$ as follows:

$$\forall\, \varphi_i \in L_2 \quad \forall\, x_i \in X \quad T \otimes I_X\left(\sum_1^n \varphi_i \otimes x_i\right) = \sum_1^n (T\varphi_i) \otimes x_i.$$

For Φ in $L_2 \otimes X$, we define, as usual, the $L_2(X)$-norm by

$$\|\Phi\|_{L_2(X)} = \left(\int \|\Phi(\omega)\|_X^2\, dm(\omega)\right)^{1/2},$$

and we define $L_2(X)$ as the completion of $L_2 \otimes X$ with respect to this norm. However, in general, the operator $T \otimes I_X$ is not bounded on $L_2 \otimes X$ equipped with this norm and does not extend to $L_2(X)$.

Actually, by a theorem of Kwapień [K2], the only spaces X for which $T \otimes I_X$ is bounded on $L_2(X)$, for all T bounded on L_2, are those which are isomorphic to a Hilbert space. More generally [K2], the only spaces X for which $T \otimes I_X$ is bounded on $L_p(X)$, for all T bounded on L_p, are those which are isomorphic to a subspace of a quotient of L_p (here $1 \leq p \leq \infty$ is fixed). We will use the following elementary and well-known fact:

Lemma 2.2. *Consider $T : L_2 \to L_2$ and a Banach space X as above. If T is positive (in the lattice sense) or if X is isomorphic to a Hilbert space, then $T \otimes I_X$ extends to a bounded operator on $L_2(X)$. We denote its norm by $\|T \otimes I_X\|$. If T is positive we have $\|T \otimes I_X\| = \|T\|$. If X is isomorphic to a Hilbert space, we have*

$$(2.11) \qquad \|T \otimes I_X\| \leq d(X, H)\|T\|.$$

Proof: (Sketch) Assume first that T is positive. Then for all sequences (φ_i) in L_2 we have

$$(2.12) \qquad \sup_i |T(\varphi_i)| \leq T(\sup_i |\varphi_i|).$$

Let Φ be an element of $L_2 \otimes X$, and let $\varphi(\omega) = \|\Phi(\omega)\|_X$. Clearly, (2.12) implies (using $\varphi(\omega) = \sup\{| < x^*, \Phi(\omega) > | \; x^* \in X^* \; \|x^*\| \leq 1\}$)

$$\|T \otimes I_X(\Phi)\| \leq T(\varphi);$$

hence

$$\|T \otimes I_X(\Phi)\|_{L_2(X)} \leq \|T\| \, \|\varphi\|_2,$$

which implies that $T \otimes I_X$ is bounded on $L_2(X)$ and $\|T \otimes I_X\| \leq \|T\|$.

Assume now that X is isometric to a Hilbert space with an orthonormal basis $\{e_i\}$.

Let $\Phi \in L_2 \otimes X$ and let $\varphi_i = < e_i, \Phi >$. We have

$$T\varphi_i = < e_i, T \otimes I_X(\Phi) >;$$

hence

$$\left(\sum |T\varphi_i|^2 \right)^{1/2} = \|T \otimes I_X(\Phi)\|_X$$

and

$$\left(\sum |\varphi_i|^2 \right)^{1/2} = \|\Phi\|_X.$$

This implies

$$\|T \otimes I_X(\Phi)\|_{L_2(X)} = \left(\sum \|T\varphi_i\|_2^2 \right)^{1/2}$$

$$\leq \|T\| \left(\sum \|\varphi_i\|_2^2 \right)^{1/2} = \|T\| \, \|\Phi\|_{L_2(X)}.$$

Hence, $T \otimes I_X$ is bounded on $L_2(X)$ and $\|T \otimes I_X\| \leq \|T\|$. If X is only assumed isomorphic to a Hilbert space, it is quite easy to complete the proof of (2.11) ∎

The reader will easily convince himself that the operator Q_1 is *not* positive (in the lattice sense). This brings us to the following:

Definition 2.3. A Banach space X is called K-convex if the operator $Q_1 \otimes I_X$ extends to a bounded operator on $L_2(X)$. In that case, we define the K-convexity constant as $K(X) = \|Q_1 \otimes I_X\|_{L_2(X) \to L_2(X)}$.

It is easy to see that K-convexity is a self-dual property, i.e. X is K-convex iff X^* is K-convex and $K(X) = K(X^*)$. Clearly ℓ_2 is K-convex and $K(\ell_2) = 1$ (cf. Lemma 2.2), but L_1 or ℓ_1 and L_∞, ℓ_∞, or c_0 are not K-convex. Actually, the class of K-convex spaces is completely elucidated by the following result from [P3] which we will not use until Chapter 11.

Theorem 2.4. *Let X be a Banach space. The following properties are equivalent.*

(i) *X is not K-convex.*
(ii) *For every $\varepsilon > 0$ and every $n \geq 1$ there is a subspace $X_n \subset X$ such that $d(X_n, \ell_1^n) < 1 + \varepsilon$.*

When (ii) holds we say that X contains ℓ_1^n's uniformly.

If a space X is isomorphic to a Hilbert space H, we have clearly (cf. Lemma 2.2) $K(X) \leq d(X, H)$. Actually, it turns out that this estimate can be improved considerably.

Theorem 2.5. *Assume X isomorphic to a Hilbert space H. Then $K(X) \leq K(1 + \mathrm{Log}\ d(X, H))$, where K is a numerical constant.*

This result is one of the crucial tools for the volume inequalities that we will prove in subsequent chapters.

To prove Theorem 2.5, we will use

Lemma 2.6. *Let $\{x_k | k \geq 0\}$ be a sequence in a Banach space B, with only finitely many non-zero terms. Assume*

$$\max_{-1 \leq \varepsilon \leq 1} \left\| \sum_0^\infty \varepsilon^k x_k \right\| \leq 1$$

and let

$$C = \max_{k \geq 0} \|x_k\|.$$

Then we have

$$\|x_1\| \leq K(1 + \operatorname{Log} C)$$

where K is a numerical constant.

Proof: We will use a classical inequality of S. Bernstein, as follows: for any integer n and any trigonometric polynomial $Q(t) = \sum_{-n \leq k \leq n} a_k e^{ikt}$ with coefficients in $\mathbf{C}$, we have

(2.13) $$\|Q'\|_\infty \leq n\|Q\|_\infty.$$

For a proof, see below.

Now, if $P(t) = \sum_{0 \leq k \leq n} x_k t^k$ is a polynomial with complex coefficients, we can define $\tilde{Q}(t) = P(\frac{1}{2}\sin t)$ and then deduce from (2.13)

$$|Q'(0)| = |\frac{1}{2}P'(0)| \leq n\|Q\|_\infty;$$

hence,

$$|P'(0)| = |x_1| \leq 2n \sup_{|t| \leq \frac{1}{2}} |P(t)|.$$

More generally, if the coefficients x_k are in a Banach space B, we can extend the previous inequality using

$$\|x_1\| = \sup\{|\zeta(x_1)| \,|\, \zeta \in B^* \;\; \|\zeta\| \leq 1\}$$

and we obtain

(2.14) $$\|x_1\| \leq 2n \sup_{|t| \leq \frac{1}{2}} \|P(t)\|.$$

We now prove Lemma 2.6. By (2.14) and the triangle inequality we have

$$\|x_1\| \leq 2n \left[\sup_{|\varepsilon| \leq \frac{1}{2}} \left\| \sum_0^\infty \varepsilon^k x_k \right\| + \sum_{k>n} 2^{-k} C \right].$$

$$\leq 2n(1 + C2^{-n})$$

Hence, choosing $n = [\operatorname{Log} C] + 1$, we obtain Lemma 2.6. (Note that if $\operatorname{Log} C$ is the logarithm in base 2, then we can take $K = 4$.) ∎

Proof of (2.13): By translation, it is sufficient to prove $|Q'(0)| \leq n\|Q\|_\infty$. We have

$$Q'(0) = i \sum_{k=-n}^{k=n} \varphi(k) a_k e^{ikt},$$

with the function φ defined by $\varphi(t) = n - |t - n|$ for $-n \leq t \leq 3n$, so that $\varphi(k) = k$ if $|k| \leq n$. We extend φ to a periodic function on $\mathbf{R}$ with period $4n$, i.e. we set $\varphi(t) = \varphi(t + 4n)$ for all t in $\mathbf{R}$. Then φ admits a Fourier series development,

$$\varphi(t) = \sum_{j\in\mathbf{Z}} c_j \exp(2\pi ijt/4n),$$

where $c_j = \int_a^{a+1} \exp(-2\pi ijt)\varphi(4nt)\,dt$ for any a in $\mathbf{R}$. Going back to the definition of $\varphi(t)$, we find $c_o = 0$, and, for $j \neq 0$ (taking $a = -\frac{1}{4}$ and $t = \frac{1}{4} + s$),

$$c_j = \int_{-1/2}^{1/2} \exp{-(2\pi ij/4)} \exp(-2\pi ijs)\,(-|s|)4n\,ds.$$

A simple computation shows that $c_j = 0$ if j is even and

$$c_j = \exp{-(2\pi ij/4)} \cdot \frac{4n}{\pi^2 j^2}$$

if j is odd.

In any case, $e^{2\pi ij/4}c_j \geq 0$ for all j; hence we can write

$$\sum |c_j| = \varphi(n) = n.$$

Therefore

$$Q'(0) = i \sum_{-n}^{n} \varphi(k)a_k = i \sum c_j Q\left(\frac{2\pi j}{4n}\right),$$

so that

$$|Q'(0)| \leq \sum |c_j|\, \|Q\|_\infty = n\|Q\|_\infty. \quad \blacksquare$$

Remark: Actually the inequality $\|Q'\|_\infty \leq 2n\|Q\|_\infty$ is enough for our purposes and is easier to prove. Indeed, let F_n be the Fejér kernel

$$F_n(t) = \sum_{|j|\leq n} \left(1 - \frac{|j|}{n+1}\right)e^{ijt}$$

and let $\psi_n(t) = 2n\, F_{n-1}(t)\sin nt$. It is well known that $\|F_n\|_1 = 1$, therefore $\|\psi_n\|_1 \leq 2n$. But it is easy to check that $Q' = -Q * \psi_n$, hence $\|Q'\|_\infty \leq 2n\|Q\|_\infty$.

Proof of Theorem 2.5: We use the operator $T(\varepsilon)$ (of Lemma 2.1) and Lemma 2.6. Recall that S—the linear span of $\{H_\alpha | \alpha \in A\}$—is dense in L_2.

For any f in $S \otimes X$ with $\|f\|_{L_2(X)} \le 1$ we consider $(T(\varepsilon) \otimes I_X)(f) = \sum_{k \ge 0} \varepsilon^k (Q_k \otimes I_X)(f)$.

By Lemma 2.2, we have

$$\|T(\varepsilon) \otimes I_X(f)\|_{L_2(X)} \le 1$$

and

$$\|Q_k \otimes I_X(f)\|_{L_2(X)} \le d(X, H).$$

Therefore, we may apply Lemma 2.6 with $C = d(X, H)$, $B = L_2(X)$ and $x_k = (Q_k \otimes I_X)(f)$ and we obtain

$$\|Q_1 \otimes I_X(f)\|_{L_2(X)} \le K(1 + \mathrm{Log}\, C).$$

By density of $S \otimes X$ in $L_2(X)$ and by homogeneity, this yields

$$K(X) = \|Q_1 \otimes I_X\|_{L_2(X) \to L_2(X)} \le K(1 + \mathrm{Log}\, C). \quad\blacksquare$$

Remark: Let $u : X \to Y$ be an operator between Banach spaces. We will say that u is K-convex if $Q_1 \otimes u$ is bounded from $L_2(X)$ to $L_2(Y)$ and we denote by $K(u)$ the norm of the corresponding operator. Assume that u factors through a Hilbert space H, i.e. $u = AB$ for some $B : X \to H$ and $A : H \to Y$. Then the same argument as above shows that

$$K(u) \le K\|u\| \left(1 + \mathrm{Log}\, \frac{\|A\|\, \|B\|}{\|u\|}\right).$$

Letting $\gamma_2(u) = \inf(\|A\|\, \|B\|)$, the infimum being over all such factorizations, we obtain

$$K(u) \le K\|u\| \left(1 + \mathrm{Log}\, \frac{\gamma_2(u)}{\|u\|}\right).$$

The notion of K-convexity yields a satisfactory duality between the space of all series $\sum g_n x_n$ with coefficients in X and the space of all series $\sum g_n x_n^*$ with coefficients in X^*. Let us make this duality more precise.

Fix an integer n. Let $\mathcal{A}_n$ be the σ-subalgebra generated by the random variables $g_1, \ldots, g_n$. We will denote by $G_n(X)$ the subspace of $L_2(\Omega, P; X)$ (simply denoted $L_2(X)$) formed by all the functions of the form $\sum_1^n g_i x_i$ with x_i in X. We will denote by $N_n(X)$ the subspace of $L_2(X)$ formed by all the $\mathcal{A}_n$-measurable functions ψ such that $\mathbf{E}(g_i \psi) = 0$ for all $i = 1, 2, \ldots$. We will denote by $G_{n*}(X)$ the space X^n equipped with the norm defined for $x = (x_1, \ldots, x_n)$ by

$$|||x||| = \inf \left\{ \left\| \sum_1^n g_i x_i + \psi \right\|_{L_2(X)} \quad \psi \in N_n(X) \right\}.$$

In other words $G_{n*}(X)$ can be identified with the quotient of $L_2(\Omega, \mathcal{A}_n; X)$ by the orthogonal of $G_n(X^*)$.

We can then summarize the announced duality as follows:

Proposition 2.7. *We have* $G_n(X^*) = (G_{n*}(X))^*$ *isometrically. Therefore, for all* $x = (x_i)$ *in* X^n

$$(2.15) \qquad |||x||| = \sup \left\{ \sum_1^n < x_i^*, x_i > \ \Big| \ \left\| \sum_1^n g_i x_i^* \right\|_{L_2(X^*)} \leq 1 \right\}.$$

Proof: Note that $G_{n*}(X)$ and $G_n(X^*)$ can be naturally identified respectively with X^n and X^{*n}, and the duality relation is the obvious one $< \sum g_i x_i^*, x > = \sum < x_i^*, x_i >$.

Let us check that the norms fit.

Let Φ be an arbitrary $\mathcal{A}_n$-measurable element of $L_2(X)$. Define

$$x_i = \mathbf{E}(g_i \Phi) \text{ for } i = 1, 2, \ldots, n.$$

Then $\psi = \Phi - \sum_1^n g_i x_i$ is clearly in $N_n(X)$ and $\mathbf{E}(< \sum_1^n g_i x_i^*, \Phi >) = \sum_1^n < x_i^*, x_i >$.

Since we have obviously

$$\left\| \sum_1^n g_i x_i^* \right\|_{L_2(X^*)} = \sup \left\{ < \sum g_i x_i^*, \Phi > \ \Big| \ \Phi \in L_2(\mathcal{A}_n; X) \ \|\Phi\|_2 \leq 1 \right\},$$

we obtain immediately

$$\left\| \sum g_i x_i^* \right\|_{G_n(X^*)} = \sup \left\{ \sum < x_i^*, x_i > \ \Big| \ |||x||| \leq 1 \right\},$$

which proves the first assertion of Proposition 2.7. The second assertion then follows immediately since $\|z\| = \sup\{< \xi, x > \ | \ |||\xi|||^* \leq 1\}$. ∎

Corollary 2.8. *Let X be a K-convex Banach space. Then with the above notation we have for all x_i in X*

$$K(X)^{-1} \left\| \sum_1^n g_i x_i \right\|_2 \leq \sup \left\{ \sum <x_i^*, x_i> \, \bigg| \, \left\| \sum g_i x_i^* \right\|_2 \leq 1 \right\}$$

$$\leq \left\| \sum g_i x_i \right\|_2 .$$

Proof: Since $Q_1 \otimes I_X(\psi) = 0$ for all ψ in $N_n(X)$, we have

$$\left\| \sum g_i x_i \right\|_2 \leq K(X) \left\| \sum g_i x_i + \psi \right\|_2 ;$$

hence

$$\leq K(X) \|\|x\|\| .$$

The first inequality then follows from Proposition 2.7. The second one is trivial. ∎

Notes and Remarks

Lemma 2.1 is a well-known classical property of the so-called Hermite semi-group (sometimes called also the Ornstein–Uhlenbeck semigroup). The kernel of the operator defined in (2.10) is often called the Mehler kernel. Lemma 2.2 is well known and elementary.

The definition of the notion of K-convexity goes back to the last remarks in the paper [MaP]. The original definition used the Rademacher functions (i.e. an i.i.d. sequence of symmetric ± 1-valued random variables) instead of Gaussian variables. Nevertheless, it was realized very early that the two definitions are equivalent and the proofs of most results can be done in either setting with minor changes. Let $K_r(X)$ be the K-convexity constant of a Banach space X, but using the Rademacher functions intead of (g_n) in Definition 2.3. Then it follows from known simple facts that $K_r(X) \leq (\pi/2)K(X)$ (see [FT] for details). Conversely, it was observed in [TJ1] that $K(X) \leq K_r(X)$.

Theorem 2.4 appeared in [P3]. Theorem 2.5 first appeared in [P7]. The simpler proof which we present via Lemma 2.6 is taken from [BM]. Proposition 2.7 is a simple observation going back to the introduction of K-convexity [MaP]. Similarly for Corollary 2.8.

Chapter 3

Ellipsoids

Historically, the ideas in this chapter originated in the 1948 work of Fritz John ([Joh]). John considered an ellipsoid of maximal volume included in the unit ball B_E of an n-dimensional normed space E. By compactness the existence of such an ellipsoid is rather obvious but John also proved its unicity. Let us denote this unique ellipsoid (often called the John ellipsoid of E) by $D_E^{\max}$. F. John also proved that

$$B_E \subset \sqrt{n} D_E^{\max},$$

which implies immediately (since $D_E^{\max} \subset B_E$ by definition)

$$d(E, \ell_2^n) \le \sqrt{n}.$$

By duality, this clearly implies the existence of a unique ellipsoid of minimum volume containing B_E. We will denote it by $D_E^{\min}$. We have necessarily by polarity from the preceding inclusion

$$n^{-1/2} D_E^{\min} \subset B_E \subset D_E^{\min}.$$

More recently, D. Lewis formulated a generalization of F. John's result which played an important role in the recent development which we want to present below.

To describe Lewis' idea, we first make precise what we mean by an ellipsoid in an n-dimensional space E. We call *ellipsoid* in E every subset $D \subset E$ which is the image of the canonical Euclidean ball $B_{\ell_2^n}$ by a linear isomorphism. Hence, $D = u(B_{\ell_2^n})$ for some invertible $u : \mathbf{R}^n \to E$.

Note that if we have an other representation $D = v(B_{\ell_2^n})$ for some isomorphism $v : \mathbf{R}^n \to E$ then necessarily $v^{-1}u$ and $u^{-1}v$ both preserve $B_{\ell_2^n}$, which means that $v^{-1}u$ is an orthogonal transformation of $\mathbf{R}^n$. Thus u is *unique modulo an orthogonal transformation*.

In analytical terms, the ellipsoid of maximal volume $D_{\max}^E$ included into B_E corresponds to an operator $u : \ell_2^n \to E$ with $\|u\| \le 1$ such that $\mathrm{vol}(u(B_{\ell_2^n}))$ is maximal.

27

It is well known that there is only one notion of *volume* on E, up to a multiplicative constant. Moreover, if E is equipped with a fixed linear basis, we may consider the determinant, denoted by $\det(u)$, of any linear map $u : \mathbf{R}^n \to E$. Then we have necessarily for some constant $c > 0$

$$\mathrm{vol}(u(B_{\ell_2^n})) = c|\det(u)|.$$

The idea of Lewis is to study the operators u which maximize $|\det(u)|$ when u runs over all operators such that $|||u||| \leq 1$, where $||| \; |||$ is now an arbitrary norm on $B(\ell_2^n, E)$ (and not only the operator norm as in John's Theorem).

Let E, F be two vector spaces.

We will denote by $\mathcal{L}(E, F)$ the space of all *linear* operators from E into F.

Let α be a norm on the space $\mathcal{L}(\mathbf{R}^n, E)$ of all linear operators from $\mathbf{R}^n$ into E. There is a well-known duality between $\mathcal{L}(\mathbf{R}^n, E)$ and $\mathcal{L}(E, \mathbf{R}^n)$ defined by

$$\forall u \in \mathcal{L}(\mathbf{R}^n, E) \qquad \forall v \in \mathcal{L}(E, \mathbf{R}^n)$$

$$< v, u >= tr(vu).$$

Let $e_1, \ldots, e_n$ be the canonical basis of $\mathbf{R}^n$.

To any u in $\mathcal{L}(\mathbf{R}^n, E)$, we can associate $x_1, \ldots, x_n$ on E defined by $x_i = u(e_i)$. Similarly, to any v in $\mathcal{L}(E, \mathbf{R}^n)$ we can associate $x_1^*, \ldots, x_n^*$ in E^* by letting $x_i^* = u^*(e_i)$. It is then easy to check that

$$(3.1) \qquad < v, u >= tr\; vu = \sum_1^n x_i^*(x_i).$$

We can introduce on $\mathcal{L}(E, \mathbf{R}^n)$ a dual norm α^* as follows:

$$(3.2) \quad \forall v : E \to \mathbf{R}^n \quad \alpha^*(v) = \sup\{tr(vT) | T : \mathbf{R}^n \to E \quad \alpha(T) \leq 1\}.$$

We now state Lewis' Theorem:

Theorem 3.1. *Let E be a vector space of dimension n and let α be a norm on $\mathcal{L}(\mathbf{R}^n, E)$. Then there is an isomorphism $u : \mathbf{R}^n \to E$ such that*

$$\alpha(u) = 1 \text{ and } \alpha^*(u^{-1}) = n.$$

Remark 3.2: Using the correspondence $u \to (u(e_1), \ldots, u(e_n))$ between $\mathcal{L}(\mathbf{R}^n, E)$ and E^n, we can reformulate Theorem 3.1 as follows.

Let α be a norm on E^n. Then there is a linear basis $(x_1, \ldots, x_n)$ of E such that the biorthogonal functionals $(x_1^*, \ldots, x_n^*)$ satisfy

$$\alpha((x_1, \ldots, x_n))\alpha^*((x_1^*, \ldots, x_1^*)) = n.$$

Proof: Let $u \to \det(u)$ be a determinant function (associated to a fixed linear basis of E). Let $K = \{u \in \mathcal{L}(\mathbf{R}^n, E) | \alpha(u) \le 1\}$. Since K is compact, the determinant attains its supremum on K; hence there exists u in K such that

$$|\det(u)| = \sup\{|\det(v)| \mid v \in K\}.$$

This implies for all T in $\mathcal{L}(\mathbf{R}^n, E)$

$$\left| \det \left(\frac{u+T}{\alpha(u+T)} \right) \right| \le |\det(u)|;$$

hence, by homogeneity,

$$(3.3) \qquad |\det(u+T)| \le |\det(u)|(\alpha(u+T))^n.$$

Clearly, $\det(u) \neq 0$, so that u is invertible, and dividing (3.3) by $|\det(u)|$ we find

$$|\det(1 + u^{-1}T)| \le \alpha(u+T)^n;$$

hence by the triangle inequality $\le (1 + \alpha(T))^n$.

Since T is arbitrary, we have for all $\varepsilon > 0$

$$(3.4) \qquad |\det(1 + \varepsilon u^{-1}T)| \le (1 + \varepsilon\alpha(T))^n.$$

But when $\varepsilon \to 0$ $\det(1 + \varepsilon u^{-1}T) = 1 + \varepsilon tr\ u^{-1}T + o(\varepsilon)$; hence (3.4) implies

$$tr\ u^{-1}T \le n\alpha(T)$$

for all $T : E \to \mathbf{R}^n$; equivalently, we have by (3.2)

$$\alpha^*(u^{-1}) \le n.$$

On the other hand, we have trivially $n = tr\ u^{-1}u \le \alpha(u)\alpha^*(u^{-1})$, hence $\alpha(u) = 1$ and $\alpha^*(u^{-1}) = n$. ∎

As an illustration, we derive a classical result of Auerbach.

Corollary 3.3.   *Let E be a normed space of dimension n. There is a basis $x_1, \ldots, x_n$ of E such that*

$$(3.5) \qquad \forall (\alpha_i) \in \mathbf{R}^n \quad \sup |\alpha_i| \le \left\| \sum_1^n \alpha_i x_i \right\| \le \sum_1^n |\alpha_i|.$$

Proof: We use Remark 3.2 with the norm $\alpha(x_1, \ldots, x_n) = \sup\|x_i\|$. By Remark 3.2, and by homogeneity there exists a basis $x_1, \ldots, x_n$ in E such that the biorthogonal functionals $x_1^*, \ldots, x_n^*$ satisfy $\max\|x_i\| = 1$ and $\sum_1^n \|x_i^*\| = n$. But since $x_i^*(x_i) = 1$, we have $\|x_i^*\| \ge 1$ for $i = 1, 2, \ldots, n$. Hence we must have $\|x_i^*\| = 1$ for $i = 1, 2, \ldots, n$. Then (3.5) follows immediately from the triangle inequality and

$$\alpha_i = x_i^* \left(\sum_{j=1}^n \alpha_j x_j \right).$$

We recall that a basis $e_1, \ldots, e_n$ for a normed space E is called 1-unconditional if we have

$$(3.6) \qquad \forall (\alpha_i) \in \mathbf{R}^n \quad \left\| \sum_1^n \alpha_i e_i \right\| = \left\| \sum_1^n |\alpha_i| e_i \right\|.$$

This implies that for any $(\alpha_i)(\beta_i)$ in $\mathbf{R}^n$ such that $|\beta_i| \le |\alpha_i|$ for all i, we have

$$\left\| \sum \beta_i e_i \right\| \le \left\| \sum \alpha_i e_i \right\|.$$

We can also derive the next result from Theorem 3.1, but we prefer to briefly reproduce a proof.

Corollary 3.4.   *Let E be an n-dimensional normed space. Let $(e_1, \ldots, e_n)$ be a 1-unconditional basis of E and let $(e_1^*, \ldots, e_n^*)$ be the dual basis of E^*. Then there are numbers $\lambda_i > 0$ such that*

$$\left\| \sum \lambda_i e_i \right\|_E = 1 \quad and \quad \left\| \sum \lambda_i^{-1} e_i^* \right\|_{E^*} = n.$$

Equivalently, we have

$$\forall (\alpha_i) \in \mathbf{R}^n \quad n^{-1} \sum_1^n |\alpha_i| \le \left\| \sum \alpha_i \lambda_i e_i \right\| \le \sup |\alpha_i|.$$

Proof: Let

$$C = \left\{ (\mu_i) \in \mathbf{R}_+^n \,\Big|\, \left\| \sum \mu_i e_i \right\| \le 1 \right\}$$

and let

$$\phi(\mu) = \prod_{i=1}^{n} \mu_i$$

for all μ in $\mathbf{R}_+^n$.

We choose λ in C such that ϕ achieves its supremum on C at λ, i.e. $\phi(\lambda) = \sup\{\phi(\mu) | \mu \in C\}$.

Then for μ in C, we have for all $\varepsilon > 0$ small enough $\lambda + \varepsilon\mu \in \mathbf{R}_+^n$ and $\phi(\lambda + \varepsilon\mu) \leq \|\sum(\lambda_i + \varepsilon\mu_i)e_i\|^n \phi(\lambda)$. Dividing by $\phi(\lambda)$, we get

$$1 + \varepsilon \sum \lambda_i^{-1}\mu_i + o(\varepsilon) \leq 1 + n\varepsilon\|\sum \mu_i e_i\| + o(\varepsilon),$$

hence

$$\sum \lambda_i^{-1}\mu_i \leq n\|\sum \mu_i e_i\| \text{ for all } \mu \text{ in } \mathbf{R}_+^n.$$

By the unconditionality condition (3.6), this is equivalent to

$$\|\sum \lambda_i^{-1}e_i^*\| \leq n.$$

On the other hand, $n \geq \|\sum \lambda_i e_i\| \, \|\sum \lambda_i^{-1}e_i^*\|$ so that we obtain the announced result.

We leave as an exercise to the reader to recognize that Corollary 3.4 is a special case of the following more abstract statement.

Corollary 3.5. *Let α be a norm on $\mathcal{L}(\mathbf{R}^n, \mathbf{R}^n)$. Let G be a compact subgroup of the group $GL(n)$ of all invertible linear maps $g : \mathbf{R}^n \to \mathbf{R}^n$. Assume that for every g in G we have*

$$\forall \, u : \mathbf{R}^n \to \mathbf{R}^n \quad \alpha(u) = \alpha(g\,u\,g^{-1}).$$

Then there is an isomorphism $u : \mathbf{R}^n \to \mathbf{R}^n$ which commutes with G such that

$$\alpha(u) = 1 \quad \text{and} \quad \alpha^*(u^{-1}) = n.$$

Proof: (Sketch.) Consider T in $\mathcal{L}(\mathbf{R}^n, \mathbf{R}^n)$. Let $\widetilde{T} = \int gTg^{-1}\,dg$ where dg is the normalized Haar measure on G. Then $\widetilde{T}$ is in $\mathcal{L}(\mathbf{R}^n, \mathbf{R}^n)$ and $\widetilde{T}$ commutes with G. We have (3.7) $\alpha(\widetilde{T}) \leq \alpha(T)$. Let then $K_G = \{u : \mathbf{R}^n \to \mathbf{R}^n | \alpha(u) \leq 1, gu = ug \, \forall g \in G\}$, and consider u in K_G such that $|\det(u)| = \sup\{|\det(w)| \, | w \in K_G\}$. We find $tr \, u^{-1}T \leq n$ for all T in K_G. Using (3.7) and the identity $tr \, u^{-1}T = tr u^{-1}\widetilde{T}$ for all $T : \mathbf{R}^n \to \mathbf{R}^n$, we obtain Corollary 3.5.

To recover Corollary 3.4 from Corollary 3.5, take for G the group of all diagonal matrices with diagonal coefficients ± 1 and take

$$\alpha(u) = \sup\{\|\sum \alpha_i u(e_i)\|_E | \sup |\alpha_i| \le 1\}.$$

We now discuss the unicity of the ellipsoid which appears in Theorem 3.1.

Proposition 3.6. In the situation of Theorem 3.1, assume that α is invariant under the orthogonal group $0(n)$, more precisely that

$$(3.8) \qquad \alpha(u) = \alpha(uT) \quad \forall\, T \in 0(n) \quad \forall\, u : \mathbf{R}^n \to E.$$

Then the operator $u : \mathbf{R}^n \to E$ such that $\alpha(u) = 1\ \alpha^*(u^{-1}) = n$ is unique modulo $0(n)$. More precisely, if an isomophism $v : \mathbf{R}^n \to E$ satisfies $\alpha(v) = 1\ \alpha^*(v^{-1}) = n$, then necesssarily $u^{-1}v$ is an orthogonal transformation.

Proof: Let v be as above and let u be as in the proof of Theorem 3.1. By the polar decomposition of $u^{-1}v$, we have

$$u^{-1}v = TA$$

with $T \in O(n)$ and A hermitian positive.

Let $\lambda_1, \ldots, \lambda_n$ be the positive eigenvalues of A. We have $v = uTA$. The definition of u implies $|\det(v)| \le |\det(u)|$; hence

$$|\det(uTA)| \le |\det(u)|$$

and since $|\det(T)| = 1$ we find

$$\Pi_1^n \lambda_i \le 1.$$

On the other hand, $v^{-1}uT = A^{-1}$ and

$$|tr\, v^{-1}uT| \le \alpha^*(v^{-1})\alpha(uT) \le n.$$

Hence

$$\sum \lambda_i^{-1} = tr\, A^{-1} \le n.$$

Clearly $\Pi\lambda_i \le 1$ and $\sum \lambda_i^{-1} \le n$ together imply $\lambda_i = 1$ for all $i \le n$ (the geometric mean of λ_i^{-1} is equal to its arithmetic mean). Therefore A is the identity and $u^{-1}v \in 0(n)$. ∎

Remark: In geometric language, the assumption (3.8) means that $\alpha(u)$ depends only on the ellipsoid $D = u(B_{\ell_2^n})$ associated to u.

Recalling how u was found in Theorem 3.1, we obtain immediately the following:

Corollary 3.7. *Let α be a norm on $\mathcal{L}(\mathbf{R}^n, \mathbf{R}^n)$ satisfying (3.8). Then, among all ellipsoids $D \subset \mathbf{R}^n$ which are of the form $D = u(B_{\ell_2^n})$ for some u with $\alpha(u) \leq 1$, there is a unique one with maximal volume. Moreover, let $D_{\max}$ be the latter ellipsoid; if $D_{\max} = u(B_{\ell_2^n})$ with $\alpha(u) \leq 1$, then necessarily $\alpha^*(u^{-1}) = n$.*

We will call this ellipsoid *the α-ellipsoid.*

When α is the operator norm from ℓ_2^n into an n-dimensional normed space E, the α-ellipsoid is nothing but the maximal volume ellipsoid $D_E^{\max} \subset B_E$.

We now recover John's Theorem in more modern formulation.

Proposition 3.8. *Let E be an n-dimensional normed space. Let $u : \ell_2^n \to E$ be an operator with $\|u\| \leq 1$ such that $u(B_{\ell_2^n})$ is the ellipsoid of maximal volume $D_E^{\max}$. Then, necessarily, $\pi_2(u^{-1}) = n^{1/2}$.*

Proof: Let $\alpha(u) = \|u\|$. By Theorem 3.1 and Corollary 3.7, we have $\alpha^*(u^{-1}) = n$. Now for $v : E \to \ell_2^n$ the dual norm $\alpha^*(v)$ is the *nuclear norm* of v, i.e. we have

$$(3.9) \qquad \alpha^*(v) = \inf\Big\{\sum_1^m \|x_i^*\|\,\|y_i\|\Big\},$$

where the infimum runs over all sets (x_i^*) in E^* and (y_i) in ℓ_2^n such that

$$\forall\, x \in E \quad v(x) = \sum_1^m x_i^*(x)y_i.$$

By Caratheodory's Theorem (since $\dim \mathcal{L}(E, \ell_2^n) = n^2$) it is sufficient to consider in (3.9) only representations with $m \leq n^2 + 1$. Thus we can find $(x_i^*), (y_i)$ as above such that

$$\forall\, x \in E \quad u^{-1}(x) = \sum_1^m x_i^*(x)y_i$$

and

$$(3.10) \qquad \sum_1^m \|x_i^*\|^2 = \sum_1^m \|y_i\|^2 = n.$$

Let $T : E \to \ell_2^m$ and $S : \ell_2^m \to \ell_2^n$ be defined by $T(x) = (x_i^*(x))$ and $S(e_i) = y_i$. Note that $ST = u^{-1}$. By (3.10), clearly $\|S\|_{HS} = n^{1/2}$. On the other hand, one immediately checks by (3.10) again that

$$\pi_2(T) \leq \Big(\sum \|x_i^*\|^2\Big)^{1/2} \leq n^{1/2}.$$

Taking into account the fact that E is of dimension n, we can compose T with the orthogonal projection onto its range, and thus we can rewrite u^{-1} as follows:

$$u^{-1} = \beta\alpha \text{ with } \alpha : E \to \ell_2^n \text{ and } \beta : \ell_2^n \to \ell_2^n$$

such that $\pi_2(\alpha) \leq n^{1/2}$ and $\|\beta\|_{HS} = n^{1/2}$.

Note that by Proposition 1.6

$$\|\alpha u\|_{HS} = \pi_2(\alpha u) \leq \pi_2(\alpha)\|u\| \leq n^{1/2}.$$

Now we observe

$$n = tr \, u^{-1} u = tr \, \beta(\alpha u);$$

hence $tr\beta(\alpha u) \geq \|\alpha u\|_{HS}\|\beta\|_{HS}$. This is the *reverse* of the Cauchy–Schwarz inequality. Therefore, we are in the case of *equality* in the Cauchy–Schwarz inequality for the scalar product $< A, B >= tr \, B^* A$. This implies (since $\|\alpha u\|_{HS} = \|\beta\|_{HS} = n^{1/2}$) that β^* must coincide with αu. Since $u^{-1} = \beta\alpha$, we have $\alpha u = \beta^{-1}$, hence $\beta^* = \beta^{-1}$. In other words, β is an orthogonal transformation so that $\|\beta\| = 1$, and therefore

$$\pi_2(u^{-1}) = \pi_2(\beta\alpha) \leq \|\beta\|\pi_2(\alpha) \leq n^{1/2}.$$

Moreover, since $\|I\|_{HS} = n^{1/2}$, we have since $I = u^{-1}u$, $n^{1/2} = \pi_2(u^{-1}u) \leq \pi_2(u^{-1})\|u\| \leq \pi_2(u^{-1})$ and we conclude as announced that $\pi_2(u^{-1}) = n^{1/2}$. ∎

Remarks:

(i) Since $\|u^{-1}\| \leq \pi_2(u^{-1}) = n^{1/2}$, we have proved in passing that the John ellipsoid $D_E^{\max}$ satisfies $B_E \subset n^{1/2}D_E^{\max}$.

(ii) Moreover, we have also proved that $\pi_2(I_E) \leq n^{1/2}$. Indeed, $\pi_2(I_E) = \pi_2(u\,u^{-1}) \leq \|u\|\pi_2(u^{-1}) = n^{1/2}$. In fact, one can show easily that $\pi_2(I_E) = n^{1/2}$ but we do not use this in the sequel.

(iii) Let $\beta(u) = n^{-1/2}\pi_2(u)$ for all $u : \ell_2^n \to E$. Then it is rather easy to show (the argument is similar to that of Proposition 3.8) that if $\beta(u) = 1$ and $\beta^*(u^{-1}) = n$, then $\|u\| = 1$. In otherwords, although $\beta(u) \leq \|u\|$, the β-ellipsoid coincides with the ellipsoid of maximal volume (associated to the operator norm of u).

In the last five chapters of this volume we will often need the following variant of Proposition 3.8.

Corollary 3.9. *Let E be an n-dimensional subspace of a Banach space X. Then there is an isomorphism $u : \ell_2^n \to E$ and an operator $v : X \to \ell_2^n$ such that $v_{|E} = u^{-1}$ and $\|u\| = 1, \pi_2(v) = n^{1/2}$.*

Proof: In the proof of Proposition 3.8, we may clearly replace the functionals (x_i^*) by their Hahn–Banach extensions, so that we may assume that $x_i^* \in X^*$ with

$$\|x_i^*\|_{X^*} = \|x_i^*\|_{E^*}.$$

Then the extended operator $v = \sum x_i^* \otimes y_i$ from X into ℓ_2^m has the desired property by the same argument as above. ∎

Recall that γ_n is the canonical Gaussian probability measure on the Euclidean space $\mathbf{R}^n$. Let E be a Banach space. For any linear operator $u : \mathbf{R}^n \to E$, we define

$$\ell(u) = \left(\int \|u(x)\|^2 \, d\gamma_n(x) \right)^{1/2}.$$

With the notation of the preceding chapter, we have

$$(3.11) \qquad \ell(u) = \left\| \sum g_i u(e_i) \right\|_{L_2(E)}$$

(here (e_i) denotes the canonical basis of $\mathbf{R}^n$).

By the rotational invariance of γ_n, the equality (3.11) remains valid when (e_i) is replaced by *any* orthonormal basis of the Euclidean space $\mathbf{R}^n$ (i.e. of ℓ_2^n). In particular, for any T isometric on ℓ_2^n we have $\ell(uT) = \ell(u)$. It follows that if $T : \ell_2^n \to \ell_2^n$ is arbitrary, then

$$(3.12) \qquad \ell(uT) \le \ell(u)\|T\|.$$

Indeed, $\ell(uT)$ is maximal on an extreme point of

$$\{T : \ell_2^n \to \ell_2^n, \|T\| \le 1\}$$

and hence on an isometry of ℓ_2^n (i.e. an orthogonal matrix).

For the composition on the other side we have trivially for any $S : E \to F$ (E, F Banach spaces)

$$\ell(Su) \le \|S\|\ell(u).$$

Therefore ℓ has the *ideal property* in the sense of [Pi1].

Let E be an arbitrary finite-dimensional space. Consider $v : E \to \ell_2^n$ and let $x_i^* = v^*(e_i)$. Then, with the notation of the preceding chapter,

$$\ell^*(v) = \sup \left\{ \sum < x_i^*, x_i > \ \middle| \ \left\| \sum_1^n g_i x_i \right\|_{L_2(E)} \le 1 \right\};$$

hence by Corollary 2.8 we have

$$\ell(v^*) = \left\| \sum g_i x_i^* \right\|_{L_2(E^*)} \le K(E^*) \ell^*(v),$$

so that we can state (recalling $K(E) = K(E^*)$):

Lemma 3.10. *For any f.d. Banach space E and any $v : E \to \ell_2^n$ we have $\ell(v^*) \le K(E) \ell^*(v)$.*

Remark. It is easy to extend the preceding result to the case of an infinite-dimensional Banach space E. Indeed, in that case let us denote by $\mathcal{F}$ the class of all f.d. subspaces of E. For F in $\mathcal{F}$, we denote by $i_F : F \to E$ the inclusion mapping.

Obviously, for all ξ in E^*, we have

$$\|\xi\| = \sup_{F \in \mathcal{F}} \|\xi_{|F}\| = \sup_{F \in \mathcal{F}} \|i_F^* \xi\|.$$

From this, it follows that for any $v : E \to \ell_2^n$

$$\ell(v^*) = \sup_{F \in \mathcal{F}} \ell((v_{i_F})^*).$$

Now by Lemma 3.10, we clearly have

$$\ell((v_{i_F})^*) \le K(F) \ell^*(v_{i_F});$$

hence *a fortiori*

$$\le K(E) \ell^*(v).$$

Therefore we obtain $\ell(v^*) \le K(E) \ell^*(v)$ in the infinite-dimensional case also.

The final statement is a recapitulation; it is a crucial tool in the sequel.

Theorem 3.11. *Let E be an n-dimensional Banach space. Then there is an isomorphism $u : \ell_2^n \to E$ such that*

$$\ell(u) = n^{1/2} \text{ and } \ell((u^{-1})^*) \le n^{1/2} K(E).$$

Equivalently, there is a basis $x_1, \ldots, x_n$ in E with biorthogonal functionals $x_1^*, \ldots, x_n^*$ such that

$$\left\| \sum g_i x_i \right\|_{L_2(E)} = n^{1/2} \text{ and } \left\| \sum_1^n g_i x_i^* \right\|_{L_2(E^*)} \le n^{1/2} K(E).$$

Moreover, we have $K(E) \le K(1 + \mathrm{Log}\,(d(E, \ell_2^n)))$ for some numerical constant K.

Proof: By Theorem 3.1, there is an isomorphism $u : \ell_2^n \to E$ such that $\ell(u) = \ell^*(u^{-1}) = n^{1/2}$. By Lemma 3.9, we have

$$\ell((u^{-1})^*) \le K(E)\ell^*(u^{-1})$$

and the above with $x_i = u(e_i)$ and $x_i^* = (u^{-1})^*(e_i)$. The upper bound for $K(E)$ was obtained in Theorem 2.5. ∎

Let X be a Banach space. We will denote by $\ell(\ell_2^n, X)$ the normed space of all operators $u : \ell_2^n \to X$ equipped with the norm ℓ.

We will use in Chapter 10 and later the following consequence of Theorem 3.1.

Corollary 3.12. *Let E be an n-dimensional subspace of a Banach space X. Then there is an isomorphism $u : \ell_2^n \to E$ and an operator $v : X \to \ell_2^n$ such that $v_{|E} = u^{-1}$ and $\ell(u) = \ell^*(v) = n^{1/2}$.*

Proof: Note that $\ell(\ell_2^n, E)$ can be considered (isometrically) as a subspace of $\ell(\ell_2^n, X)$. By Theorem 3.1, we have $u : \ell_2^n \to E$ such that $\ell(u) = \ell^*(u^{-1}) = n^{1/2}$.

Consider the linear form $\xi(w) = tr\,u^{-1}w$ defined for all w in $\ell(\ell_2^n, E)$. Then $\|\xi\| = \ell^*(u^{-1})$. By the Hahn–Banach Theorem, we can extend ξ to a linear form $\tilde{\xi}$ on $\ell(\ell_2^n, X)$ with the same norm. Clearly there is an operator $v : X \to \ell_2^n$ such that $\tilde{\xi}(w) = tr\,vw$ for all $w : \ell_2^n \to X$. Since $\tilde{\xi}$ extends ξ, v must extend u^{-1} and we have $\ell^*(v) = \|\tilde{\xi}\|$. Since $\|\tilde{\xi}\| = \|\xi\|$, we conclude $\ell^*(v) = \ell^*(u^{-1})$. ∎

We now define the ℓ-norm in the infinite dimensional case. Consider a (possibly infinite dimensional) Hilbert space H. Let $u : H \to X$ be an operator with values in a Banach space X. For any f.d. subspace $S \subset H$ with an orthonormal basis $(e_1, \ldots, e_n)$ we may identify S with ℓ_2^n so that $\ell(u_{|S})$ is unambiguously defined. Note that by (3.12) it does not depend on the choice of the basis $e_1, \ldots, e_n$ of S with which we identify S and ℓ_2^n. Then we can define the (possibly infinite) number

$$(3.13) \qquad \ell(u) = \sup\{\ell(u_{|S}) \big| S \subset H, \dim S < \infty\}.$$

(If H is finite dimensional, this coincides by (3.12) with the preceding definition (3.11).)

The operators u for which $\ell(u) < \infty$ will be called ℓ-operators and the set of all such operators will be denoted by $\ell(H, X)$. This set becomes a Banach space when equipped with the norm ℓ. Note that (3.13) may be equivalently rewritten (using (3.12)) as

$$\ell(u) = \sup\{\ell(uT) \big| n \geq 1, T : \ell_2^n \to H, \|T\| \leq 1\}.$$

With this alternate definition, it is clear that the *ideal property* holds, i.e. for all Hilbert spaces H, H_1, all Banach spaces X, X_1 and all operators $S : X \to X_1$ and $T : H_1 \to H$ we have

$$\ell(SuT) \leq \|S\|\ell(u)\|T\|.$$

To give some examples of ℓ-operators, it is easy to check using (1.15) that every 2-summing operator $u : H \to X$ is an ℓ-operator and that we have

$$(3.14) \qquad\qquad \ell(u) \leq \pi_2(u).$$

(Indeed, consider $S \subset H$ with orthonormal basis $e_1, \ldots, e_n$ and let $\phi = \sum g_i e_i$. By (1.15) we have $\ell(u_{|S}) = \| \sum g_i u(e_i) \|_{L_2(X)} \leq \pi_2(u)$.)

However, in general, ℓ-operators are more general than 2-summing operators. (To see this, it is enough to show that the norms π_2 and ℓ are not equivalent. For that purpose, consider the inclusion mapping $j_n : \ell_2^n \to \ell_\infty^n$, then it is easy to check that $\pi_2(j_n) \geq n^{1/2}$ while $\ell(j_n)$ is $O((\text{Log } n)^{1/2})$, so that π_2 and ℓ are not equivalent norms on the space of finite rank operators from ℓ_2 into ℓ_∞.) Nevertheless, if we restrict ourselves to Hilbert spaces, we recover Hilbert Schmidt operators.

Proposition 3.13. *Let H_1, H_2 be Hilbert spaces and let X be a Banach space isomorphic to H_2. Then every ℓ-operator $u : H_1 \to X$ is 2-summing and we have*

$$(3.15) \qquad \pi_2(u) \leq d(X, H_2)\ell(u).$$

Moreover,

$$(3.16) \qquad \left(\sum a_n(u)^2\right)^{1/2} \leq d(X, H_2)\ell(u).$$

Proof: Let $x_1, \ldots, x_n$ in H_1 be such that

$$(3.17) \qquad \sup\left\{\left(\left|\sum \xi(x_i)\right|^2\right)^{1/2}\middle|\xi \in B_{H_1}\right\} \leq 1.$$

Let $e_1, \ldots, e_n$ be the canonical basis of ℓ_2^n. Consider $T : \ell_2^n \to H_1$ defined by $T(e_i) = x_i$. Let $S : X \to H_1$ be an isomorphism. Let $h_i = Su(x_i) \in H_2$. Clearly, using the inner product of H_2, we have

$$(3.18) \qquad \ell(SuT) = \left\|\sum g_i h_i\right\|_{L_2(H_2)} = \left(\sum \|h_i\|^2\right)^{1/2}.$$

Therefore,

$$\begin{aligned}\left(\sum \|ux_i\|^2\right)^{1/2} &\leq \|S^{-1}\|\left(\sum \|h_i\|^2\right)^{1/2}\\ &\leq \|S^{-1}\|\ell(SuT)\\ &\leq \|S^{-1}\|\,\|S\|\,\ell(u)\|T\|.\end{aligned}$$

By (3.17) we have $\|T\| \leq 1$; hence we conclude

$$\left(\sum \|ux_i\|^2\right)^{1/2} \leq d(X, H_2)\ell(u),$$

and this implies (3.15) by homogeneity. Finally, (3.16) follows immediately from (3.15) and Corollary 1.9.

Remark 3.14: Using Remark 1.5 the reader can check rather easily that Urysohn's inequality has the following consequence: for all operators $u : \ell_2^n \to X$ with values in an arbitrary Banach space, we have

$$\left(\frac{\text{vol}(u^*(B_{X^*}))}{\text{vol}(B_{\ell_2^n})}\right)^{1/n} \leq \ell(u).$$

From this and Theorem 3.11, it is easy to deduce that $K(E) = 1$ iff E is isometric to ℓ_2^n.

Notes and Remarks

Theorem 3.1 and Corollary 3.7 are Lewis' generalization of F. John's Theorem [Joh]. These results appeared in [L].

Corollary 3.3 is a well-known classical statement due to Auerbach, and Corollary 3.4 is a well-known fact due to Lozanovskii [Lo]. Corollary 3.5 is a standard averaging argument. Concerning Theorem 3.8, the fact that $\pi_2(I_E) = n^{1/2}$ for any n-dimensional space E is due to Garling and Gordon [GaG]. For a simple proof—due to Kwapień— using the Pietsch factorization Theorem, cf. [P2].

Corollary 3.9 is an immediate consequence of the Garling–Gordon Theorem and the Hahn–Banach Theorem. On the other hand, the fact that John's ellipsoid has the properties given in Theorem 3.8 has been known to many people for a long time. Lemma 3.10 is nothing but a reformulation of the (Gaussian) definition of K-convexity, while Theorem 3.11 and its corollary are merely a combination of Lewis' Theorem 3.1 with Lemma 3.10. However, we should point out that these results were used in this way and in a similar context for the first time by Figiel and Tomczak–Jaegermann in [FT]. They were inspired by previous work of Lewis (see [FT]) in the case of Banach lattices. Proposition 3.13 is a well-known result due to Pietsch [Pi1] who first introduced and studied the p-absolutely summing operators which had been considered in the cases $p = 1$ and $p = 2$ by Grothendieck under a different name.

Chapter 4
Dvoretzky's Theorem

In this chapter we prove the fundamental theorem of Dvoretzky on the almost spherical (or rather *ellipsoidal*) sections of convex bodies. Although this result is not directly related to volume estimates, it has many points of contact with the topics of the other chapters, and we feel it belongs in this volume. We first state Dvoretzky's Theorem.

Theorem 4.1. *For each $\varepsilon > 0$ and n, there is a number $N(\varepsilon, n)$ such that every Banach space E of dimension at least $N(\varepsilon, n)$ contains an n-dimensional subspace F such that $d(F, \ell_2^n) \leq 1 + \varepsilon$.*

For infinite-dimensional Banach spaces, we note the following immediate consequence.

Corollary 4.2. *Let E be an infinite-dimensional Banach space. Then, for each $\varepsilon > 0$, there is a sequence $\{E_n\}$ of subspaces of E such that $d(E_n, \ell_2^n) \leq 1 + \varepsilon$ for all n. (In other words, we often say that E contains ℓ_2^n's $(1 + \varepsilon)$-uniformly.)*

But it is actually the finite-dimensional case which is the most interesting for us. In that case we will prove the following refinement of Theorem 4.1.

Theorem 4.3. *For each $\varepsilon > 0$, there is a number $\eta(\varepsilon) > 0$ with the following property. Let E be an f.d. Banach space of dimension N. Then E contains a subspace $F \subset E$ of dimension $n = [\eta(\varepsilon) \operatorname{Log} N]$ such that $d(F, \ell_2^n) \leq 1 + \varepsilon$.*

Remarks:

(i) In geometric terms, the above theorem says that if
$$n = [\eta(\varepsilon) \operatorname{Log} N]$$
then any ball $B \subset \mathbf{R}^N$ admits a *section* which is $(1+\varepsilon)$-equivalent to an ellipsoid, i.e. there is a subspace $F \subset \mathbf{R}^N$ with $\dim F = n$ and an ellipsoid $D \subset F$ such that
$$D \subset F \cap B \subset (1 + \varepsilon)D.$$

41

Clearly, by duality, a similar statement holds for *projections* of
B: There is an n-dimensional projection of B which is $(1 + \varepsilon)$-
equivalent to an ellipsoid.

(ii) In other words, every normed space E of dimension N admits,
for $n = [\eta(\varepsilon) \operatorname{Log} N]$, an n-dimensional *quotient* space which is
$(1 + \varepsilon)$-isomorphic to ℓ_2^n. This trivial *dualization* applies also
to Theorem 4.1 and Corollary 4.2: every infinite dimensional
Banach space E admits, for each $\varepsilon > 0$, a sequence $\{E_n\}$ of
quotient spaces such that $d(E_n, \ell_2^n) \leq 1 + \varepsilon$ for all n.

(iii) Clearly, Theorem 4.3 implies Theorem 4.1 with

$$N(\varepsilon, n) = \exp\left(\frac{n}{\eta(\varepsilon)}\right).$$

The original proof of Theorem 4.1 appeared in [Dv]. For subsequent
proofs cf. [M6], [Sz], [F]. In this chapter we follow the *measure-
theoretical* point of view of Milman's proof [M6], but we use Gaussian
measures instead of surface measures on Euclidean spheres.

We will prove the following theorem, which may be viewed as a
Gaussian reformulation of Dvoretzky's Theorem.

Theorem 4.4.   *For each $\varepsilon > 0$ there is a number $\eta_1(\varepsilon) > 0$ for
which the following statement holds. As before, let (g_k) be an i.i.d. sequence of Gaussian normal random variables on some probablity space
$(\Omega, \mathcal{A}, \mathbf{P})$. Let E be a Banach space and let (z_k) be a sequence of elements of E. We assume that the series*

$$(4.1) \qquad\qquad X = \sum_{k=1}^{\infty} g_k z_k$$

converges in $L_1(\Omega, \mathcal{A}, \mathbf{P}; E)$. Let

$$\sigma(X) = \sup\left\{ \left(\sum |\zeta(z_k)|^2\right)^{1/2} \,\Big|\, \zeta \varepsilon E^* \quad \|\zeta\| \leq 1 \right\}$$

and

$$d(X) = (\mathbf{E}\|X\|/\sigma(X))^2.$$

Then, for each $\varepsilon > 0, E$ contains a subspace F of dimension

$$n = [\eta_1(\varepsilon)\, d(X)]$$

which is $(1 + \varepsilon)$-isomorphic to ℓ_2^n.

Note: We have $\mathbf{E}|\zeta(X)| = \mathbf{E}|g_1| \left(\sum |\zeta(z_k)|^2\right)^{1/2}$ by (2.5), and $\mathbf{E}|g_1| = (2/\pi)^{1/2}$, hence $\sigma(X) \leq (\pi/2)^{1/2}\mathbf{E}\|X\|$.

Remark 4.5: The preceding theorem is proved in detail below, but we wish to give first here a general idea of the proof: Let $X_1, X_2, \ldots, X_n$ be i.i.d. copies of the variable X on some probability space $(\Omega, \mathcal{A}, \mathbf{P})$. Let $\Omega(n, \varepsilon)$ be the set of all ω in Ω such that

$$\forall\,(\alpha_i) \in \mathbf{R}^n \quad (1+\varepsilon)^{-1/2}\left(\sum |\alpha_i|^2\right)^{1/2} \leq \left\|\sum_{i=1}^n \alpha_i X_i(\omega)(\mathbf{E}\|X\|)^{-1}\right\|$$
$$\leq (1+\varepsilon)^{1/2}\left(\sum |\alpha_i|^2\right)^{1/2}.$$

We will show below that $\mathbf{P}(\Omega(n, \varepsilon)) > 0$, provided that n is not too large and precisely provided $n \leq [\eta_1(\varepsilon)\,d(X)]$. This clearly yields Theorem 4.4.

Remarks:

(i) We call $d(X)$ briefly the *dimension* of X (perhaps the term *concentration dimension* would be more appropriate). Note, however, that this notion of dimension depends not only on X but also on the norm of the Banach space E into which X takes its values. Also note that $d(X)$ is a real number and not necessarily an integer.

(ii) In [P1], we have defined $d(X)$ as the ratio $\mathbf{E}\|X\|^2/\sigma(X)^2$. Note that by Corollary 4.9 below this is equivalent to the above definition.

(iii) If E is f.d. with $\dim E = N$, we know (cf. Chapter 3) that $\pi_2(I_E) \leq N^{1/2}$. This implies, using (1.15) that

$$\mathbf{E}\|X\| \leq (\mathbf{E}\|X\|^2)^{1/2} \leq N^{1/2} \sup\left\{(\mathbf{E}|\zeta(X)|^2)^{1/2}\big|\zeta \in B_{E^*}\right\}.$$

Therefore we have $d(X) \leq N = \dim E$.

(iv) The above Theorem 4.4 reduces the task of proving Dvoretzky's Theorem for E to that of exhibiting E-valued Gaussian variables X with large dimension $d(X)$. More precisely, let us denote by $n_\varepsilon(X)$ the largest integer n such that there is an n-dimensional subspace $F \subset E$ satisfying $d(F, \ell_2^n) \leq 1 + \varepsilon$.

Also, let $\delta(E) = \sup d(X)$ where the supremum runs over all E-valued Gaussian variables X as in (4.1). Then by Theorem 4.4 we have for some function $\eta_2(\varepsilon) > 0$

$$(4.2) \qquad n_\varepsilon(E) \geq \eta_2(\varepsilon)\delta(E).$$

There is also a converse estimate which shows that this *measure theoretic approach* to Dvoretzky's Theorem is somewhat sharp. This is much easier to check as follows.

Proposition 4.6.

$$\delta(E)\left(\frac{\pi}{2}\right)(1+\varepsilon)^2 \geq n_\varepsilon(E).$$

Proof: Let $F \subset E$ be such that $d(F, \ell_2^n) \leq 1+\varepsilon$. Let $T : \ell_2^n \to F$ be such that $\|T\|\,\|T^{-1}\| \leq 1+\varepsilon$. We define

$$X = \sum_{i=1}^n g_i T(e_i).$$

Then $\sigma(X) = \|T\|$ and

$$\|T^{-1}\|\mathbf{E}\|X\| \geq \mathbf{E}\|T^{-1}(X)\| = \mathbf{E}\left(\sum_1^n |g_i|^2\right)^{1/2}$$

$$\geq \left(\sum_1^n (\mathbf{E}|g_i|)^2\right)^{1/2} = n^{1/2}(2/\pi)^{1/2}.$$

This implies $d(X) \geq (2/\pi)n(\|T\|\,\|T^{-1}\|)^{-2} \geq (2/\pi)n(1+\varepsilon)^{-2}$, which yields Proposition 4.6. ∎

For the proof of Theorem 4.4, the following estimation of the deviation of $\|X\|$ from its mean will be crucial. In Milman's terminology, this is a *concentration of measure* phenomenon.

Theorem 4.7. *Let X be as in Theorem 4.4. We have*

(4.3) $\forall\, t > 0$ $\mathbf{P}\{|\|X\| - \mathbf{E}\|X\|| > t\} \leq 2\exp(-Kt^2/\sigma(X)^2),$

where K is a numerical constant ($K = 2\pi^{-2}$ in the proof below).

Proof: We may clearly assume that E is f.d. and that the series (4.1) defining X is actually a finite sum,

$$X = \sum_1^m g_k z_k \qquad (z_k \in E).$$

We may also assume non-degeneracy so that the distribution of X is equivalent to the Lebesgue measure on E. Then let $u : \ell_2^m \to E$ be the linear operator defined by

$$\forall\, x \in \mathbf{R}^m \quad u(x) = \sum_1^m x_k z_k.$$

Let γ_m be the canonical Gaussian probability measure on $\mathbf{R}^m$. Then (4.3) reduces to the following inequality
(4.4)

$$\gamma_m \left\{ x \in \mathbf{R}^m \,\Big|\, \big|\|u(x)\| - \int \|u(x)\|\, d\gamma_m\big| > t \right\} \le 2 \exp\left(-\frac{K\, t^2}{\|u\|^2} \right).$$

Indeed, it is easy to check that $\sigma(X) = \|u^*\| = \|u\|$.

To prove (4.4), we will work on $\mathbf{R}^m \times \mathbf{R}^m$ equipped with $\gamma_m \times \gamma_m$ and we denote by (x, y) a generic point of $\mathbf{R}^m \times \mathbf{R}^m$.

For θ in $[0, 2\pi]$, let then

$$x(\theta) = x \sin\theta + y \cos\theta,$$
$$x'(\theta) = x \cos\theta - y \sin\theta.$$

We also denote

$$F(x) = \|u(x)\|.$$

Note that $F : \mathbf{R}^m \to \mathbf{R}$ is Lipschitzian, hence admits a derivative (or a gradient) at almost every point x in $\mathbf{R}^m$. Let x, y be fixed for the moment and assume that F is differentiable in $x(\theta)$ for almost every θ. Then we can write

$$F(x) - F(y) = F(x(\pi/2)) - F(x(0))$$

(4.5)
$$= \int_0^{\pi/2} \frac{d}{d\theta} F(x(\theta))\, d\theta$$

$$= \int_0^{\pi/2} F'(x(\theta)) \cdot x'(\theta)\, d\theta.$$

(Here we use the notation $F(x + h) = F(x) + F'(x) \cdot h + o(\|h\|)$ when $h \to 0$, so that $F'(x)$ is viewed as a linear form on $\mathbf{R}^m$.)

Let λ be a real number and let $\Phi_\lambda : \mathbf{R} \to \mathbf{R}$ be the function $\Phi_\lambda(t) = \exp(\lambda t)$ for all t in $\mathbf{R}$. By convexity, we deduce from (4.5)

(4.6) $$\Phi_\lambda(F(x) - F(y)) \le \frac{2}{\pi} \int_0^{\pi/2} \Phi_\lambda\left(\frac{\pi}{2} F'(x(\theta)) x'(\theta) \right) d\theta.$$

We wish now to integrate (4.6) in x, y with respect to $\gamma_m \times \gamma_m$. Note that (by Fubini's Theorem) since $F'(x)$ exists for a.e. x, for a.e. couple (x, y) the derivative $F'(x(\theta))$ exists for a.e. θ. Integrating (4.6), we obtain

$$(4.7) \quad \begin{aligned} \iint & \Phi_\lambda(F(x) - F(y)) \, d\gamma_m(x) \, d\gamma_m(y) \\ & \leq \frac{2}{\pi} \int_0^{\pi/2} \iint \Phi_\lambda\left(\frac{\pi}{2} F'(x(\theta)) \cdot x'(\theta)\right) d\theta \, d\gamma_m(x) \, d\gamma_m(y). \end{aligned}$$

By convexity, we have

$$\begin{aligned} \int \Phi_\lambda & \left(F(x) - \int F(y) \, d\gamma_m(y)\right) d\gamma_m(x) \\ & \leq \iint \Phi_\lambda(F(x) - F(y)) \, d\gamma_m \, d\gamma_m. \end{aligned}$$

On the other hand, we observe (this is the crucial point) that for each θ the measure $\gamma_m \times \gamma_m$ is invariant under the mapping $(x, y) \to (x(\theta), x'(\theta))$. Indeed, this follows from the *rotational* invariance of Gaussian measures. Therefore the right-hand side of (4.7) is equal to

$$\iint \Phi_\lambda\left(\frac{\pi}{2} F'(x) \cdot y\right) d\gamma_m(x) \, d\gamma_m(y).$$

Let us denote by $\| \ \|_2$ the Euclidean norm on $\mathbf{R}^m$. It is well known that for any ζ in $\mathbf{R}^m$

$$\int \exp(\zeta(y)) \, d\gamma_m(y) = \exp\frac{1}{2}\|\zeta\|_2^2,$$

hence

$$\int \exp\left(\frac{\pi}{2}\lambda F'(x)y\right) d\gamma_m(y) = \exp\frac{1}{2}\left(\frac{\pi}{2}\right)^2 \|F'(x)\|_2^2 \lambda^2.$$

But since $F(x) = \|u(x)\|$ we have

$$|F(x) - F(y)| \leq \|u\| \, \|x - y\|_2 \quad \forall \, x, y \in \mathbf{R}^m;$$

hence

$$\|F'(x)\|_2 \leq \|u\| \quad \text{for a. e. } x \text{ in } \mathbf{R}^m.$$

Therefore we deduce finally from (4.7) that $\forall \, \lambda \in \mathbf{R}$

$$\int \Phi_\lambda\left(F(x) - \int F(y) \, d\gamma_m(y)\right) d\gamma_m(x) \leq \exp\frac{1}{2}\left(\frac{\pi}{2}\right)^2 \|u\|^2 \lambda^2.$$

By Markov's inequality, we have *a fortiori* for all $\lambda > 0$

$$\gamma_m \left\{ x \Big| F(x) - \int F \, d\gamma_m > t \right\} \leq \exp \left[\frac{1}{2} \left(\frac{\pi}{2} \right)^2 \|u\|^2 \lambda^2 - \lambda t \right];$$

choosing $\lambda = \left(\frac{\pi}{2} \|u\| \right)^{-2} t$, we find

$$\gamma_m \left\{ x \Big| F(x) - \int F \, d\gamma_m > t \right\} \leq \exp -\frac{1}{2} \left(\frac{\pi}{2} \|u\| \right)^{-2} t^2.$$

Clearly, the same inequality holds for $-F$ also, so that we finally obtain the announced inequality (4.4) and hence also (4.3). ∎

Remark 4.8: The preceding proof actually establishes the following: for any Lipschitzian function $F : \mathbf{R}^m \to \mathbf{R}$ such that

$$\forall \, x, y \in \mathbf{R}^m \quad |F(x) - F(y)| \leq \sigma \|x - y\|_2$$

we have

$$\gamma_m \left\{ x \in \mathbf{R}^m \Big| \, \Big| F(x) - \int F \, d\gamma_m \Big| > t \right\} \leq 2 \exp -K t^2 \sigma^{-2}$$

for a numerical constant K.

The same argument can easily be adapted to other rotational invariant measures than γ_m. For instance, let σ_m be the normalized surface measure on the Euclidean unit sphere S of $\mathbf{R}^m$. Then a similar argument yields

$$\sigma_m \left\{ x \in S \Big| \, \Big| F(x) - \int F d\sigma_m \Big| > t \right\} \leq 2 \exp \left(-K' t^2 n \sigma^{-2} \right),$$

where K' is a numerical constant and σ is the Lipschitz norm of F restricted to S.

Using the latter instead of (4.3), it is possible to prove the following claim. Given $u : \ell_2^m \to E$, there is an orthonormal basis $x_1, \ldots, x_m$ of ℓ_2^m such that the n first vectors $x_1, \ldots, x_n$ satisfy

$$\forall \, \alpha \in \mathbf{R}^n \quad (1 + \varepsilon)^{-1/2} \left(\sum_1^n |\alpha_i|^2 \right)^{1/2} \leq \frac{1}{M} \Big\| \sum_1^n \alpha_i x_i \Big\|$$

$$\leq (1 + \varepsilon)^{1/2} \left(\sum_1^n |\alpha_i|^2 \right)^{1/2},$$

where

$$M = \int \|u(x)\| \, d\sigma_m(x)$$

and where

$$n = [\eta_1'(\varepsilon) M^2 \|u\|^{-2}].$$

This should be compared with Remark 4.5. Indeed, working with the normalized Haar measure on the orthogonal group $O(m)$, we can choose the orthonormal basis $x_1, \ldots, x_m$ at *random* and show using the above estimates that with positive probability, we have the preceding claim.

Corollary 4.9. *Let X be as in Theorem 4.7. We have for all $p > 1$*

$$\mathbf{E}\|X\| \le (\mathbf{E}\|X\|^p)^{1/p} \le K(p)\mathbf{E}\|X\|,$$

where $K(p)$ is a constant depending only on p.

Proof: Note that (as already observed above)

$$\mathbf{E}|\zeta(X)| = \left(\frac{2}{\pi}\right)^{1/2} (\mathbf{E}|\zeta(X)|^2)^{1/2};$$

hence $\sigma(X) \le (\pi/2)^{1/2}\mathbf{E}\|X\|$.

Assume for simplicity that $\mathbf{E}\|X\| = 1$. Then by Theorem 4.7 we have

$$(4.8) \qquad \mathbf{P}\{\|X\| > t + 1\} \le 2 \exp - \left(Kt^2\left(\frac{2}{\pi}\right)\right).$$

But

$$\mathbf{E}\|X\|^p = \int_0^\infty pt^{p-1}\mathbf{P}\{\|X\| > t\} \, dt;$$

hence (4.8) implies

$$\mathbf{E}\|X\|^p \le K(p)^p$$

for some constant $K(p)$ depending only on p. ∎

To prove Theorem 4.4 we will need two elementary (but rather technical) lemmas which will allow us to *discretize* the problem of finding an ℓ_2^n-subspace.

Lemma 4.10. *Let $||| \; |||$ be any norm on $\mathbf{R}^n$ with unit ball B and unit sphere S. Let $\delta > 0$. There is a subset $A \subset S$ which is a δ-net of S with respect to the norm $||| \; |||$ with cardinality*

$$\operatorname{card}(A) \leq \left(1 + \frac{2}{\delta}\right)^n.$$

Proof: Let $(y_i)_{i \leq m}$ be a maximal subset of S such that $|||y_i - y_j||| \geq \delta$ for all $i \neq j$. Clearly, by maximality, $(y_i)_{i \leq m}$ is a δ-net of S. To majorize m, we note that the balls $y_i + (\delta/2)B$ are disjoint and included into $(1 + \delta/2)B$. Therefore,

$$\sum_{i \leq m} \operatorname{vol}\left(y_i + \frac{\delta}{2}B\right) \leq \operatorname{vol}\left(\left(1 + \frac{\delta}{2}\right)B\right) = \left(1 + \frac{\delta}{2}\right)^n \operatorname{vol}(B);$$

hence

$$m\left(\frac{\delta}{2}\right)^n \operatorname{vol}(B) \leq \left(1 + \frac{\delta}{2}\right)^n \operatorname{vol}(B),$$

and this implies $m \leq (1 + 2/\delta)^n$ as announced.

The next lemma shows that in the event $\Omega(n, \varepsilon)$ (cf. Remark 4.5), it is enough to consider coefficients $\alpha = (\alpha_i)$ in a δ-net of the Euclidean sphere for δ sufficiently small.

Lemma 4.11. *For each $\varepsilon > 0$, there is a $\delta = \delta(\varepsilon), 0 < \delta < 1$, with the following property. Let n be any integer and let $||| \; |||$ be a norm on $\mathbf{R}^n$. Let A be a δ-net in the unit sphere of $(\mathbf{R}^n, ||| \; |||)$ and let $x_1, \ldots, x_n$ be elements of a Banach space B. If*

$$\forall \, \alpha \in A \quad 1 - \delta \leq \left\|\sum_1^n \alpha_i x_i\right\| \leq 1 + \delta,$$

then

$$\forall \, \alpha \in \mathbf{R}^n \quad (1 + \varepsilon)^{-1/2}|||\alpha||| \leq \left\|\sum_1^n \alpha_i x_i\right\| \leq (1 + \varepsilon)^{-1/2}|||\alpha|||.$$

In particular, let F be the subspace generated by $x_1, \ldots, x_n$, then F is $(1 + \varepsilon)$-isomorphic to $(\mathbf{R}^n, ||| \cdot |||)$.

Proof: Assume $|||\alpha||| = 1$. There is y° in S such that $|||\alpha - y^\circ||| \leq \delta$ hence $\alpha = y^\circ + \lambda_1 \alpha'$ with $|\lambda_1| \leq \delta$ and $|||\alpha'||| = 1$. Continuing this process we obtain $\alpha = y^\circ + \lambda_1 y^1 + \lambda_2 y^2 + \ldots$ with $y^j \in S$ and $|\lambda_j| \leq \delta^j$.

It follows that

$$\left\|\sum \alpha_i x_i\right\| \le \sum_{j \ge 0} \delta^j \left\|\sum_i y_i^j x_i\right\|$$
$$\le (1+\delta)(1-\delta)^{-1}.$$

Similarly,

$$\left\|\sum \alpha_i x_i\right\| \ge \left\|\sum y_i^\circ x_i\right\| - \delta(1+\delta)(1-\delta)^{-1}$$
$$\ge 1 - \delta - \delta(1+\delta)(1-\delta)^{-1} = (1-3\delta)(1-\delta)^{-1}.$$

Hence, if $\delta = \delta(\varepsilon)$ is chosen small enough so that

$$(1-3\delta)(1-\delta)^{-1} \ge (1+\varepsilon)^{-1/2} \text{ and } (1+\delta)(1-\delta)^{-1} \le (1+\varepsilon)^{1/2}$$

we obtain (by homogeneity) the announced result. Note that we can find a suitable $\delta > 0$ depending only on ε (and *independent of N*).

Remark 4.12: Let p be an arbitrary semi-norm on $\mathbf{R}^n$. Then the proof of Lemma 4.11 shows clearly that

$$\sup\{p(\alpha)\big|\alpha \in \mathbf{R}^n \quad |||\alpha||| \le 1\} \le (1-\delta)^{-1}\sup\{p(\alpha)\big|\alpha \in A\}.$$

Proof of Theorem 4.4: Let $X_1, X_2, \ldots, X_n$ be i.i.d. copies of X. We proceed as indicated above in Remark 4.5. We apply the two preceding lemmas with $|||\alpha||| = \left(\sum_1^n |\alpha_i|^2\right)^{1/2}$ (the Euclidean norm on $\mathbf{R}^n$). Assume $\sum |\alpha_i|^2 = 1$. Then the r.v. $\sum_1^n \alpha_i X_i$ has the same distribution as X. This classical fact follows from (and is a generalization of) (2.3).

Therefore we have $\mathbf{E}\|\sum \alpha_i X_i\| = \mathbf{E}\|X\|$ and by (4.3)

$$\forall \delta > 0 \quad \mathbf{P}\left\{\left|\left\|\sum \alpha_i X_i\right\| - \mathbf{E}\|X\|\right| > \delta \mathbf{E}\|X\|\right\} \le 2\exp(-K\delta^2 d(X)).$$

Let $M = \mathbf{E}\|X\|$ and let A be as in Lemma 4.10 and 4.11. The preceding inequality implies
(4.9)
$$\mathbf{P}\left\{\exists \alpha \in A \left|\left\|\sum \alpha_i X_i\right\|M^{-1} - 1\right| > \delta\right\} \le 2|A|\exp(-K\delta^2 d(X)).$$

Hence, by Lemma 4.10,

$$\le 2\exp\left\{\frac{2n}{\delta} - K\delta^2 d(X)\right\}.$$

Now we choose $\delta = \delta(\varepsilon)$, the function of ε given by Lemma 4.11. This yields the desired result. Indeed, observe that if

$$(4.10) \qquad \frac{2n}{\delta} \leq \frac{1}{2} K \delta^2 d(X)$$

the probability (4.9) is not greater than $2 \exp -(1/2) K \delta^2 d(X)$; moreover, we can always assume that $d(X)$ is large enough (say $d(X) \geq 4(K\delta^2)^{-1}$), otherwise there is nothing to prove. Hence we can assume the right side of (4.9) < 1. We then obtain that with *positive* probability we have

$$\forall \, \alpha \in A \quad \left| M^{-1} \left\| \sum \alpha_i X_i(\omega) \right\| - 1 \right| \leq \delta.$$

By Lemma 4.11 (recall we chose $\delta = \delta(\varepsilon)$) we conclude that for all α in the sphere of ℓ_2^n we have

$$(1+\varepsilon)^{-1/2} \leq M^{-1} \left\| \sum_1^n \alpha_i X_i \right\| \leq (1+\varepsilon)^{1/2},$$

and by homogeneity this implies (with the notation of Remark 4.5) that $\mathbf{P}(\Omega(n,\varepsilon)) > 0$.

Hence if F_ω is the span of $(X_1(\omega), \ldots, X_n(\omega))$ and (recalling (4.10)) if $n = [4^{-1} K \delta(\varepsilon)^3 d(X)]$, we have proved that with positive probability we have $d(F_\omega, \ell_2^n) \leq 1 + \varepsilon$. ∎

We now turn to the proof of Theorem 4.3. For this we will use the following variant of a classical lemma of Dvoretzky–Rogers:

Lemma 4.13. *Let E be an N-dimensional space. Let $\overline{N} = [N/2]$. Then there are $\overline{N}$ elements $(x_k)_{k \leq \overline{N}}$ in E satisfying*

$$(4.11) \qquad \forall \, \alpha \in \mathbf{R}^{\overline{N}} \quad \left\| \sum \alpha_k x_k \right\| \leq \left(\sum |\alpha_k|^2 \right)^{1/2}$$

and

$$(4.12) \qquad \|x_k\| \geq \tfrac{1}{2} \text{ for all } k \leq \overline{N}.$$

Proof: Let $\alpha(u) = \|u\|$. By Theorem 3.1, there is an isomorphism $u : \ell_2^N \to E$ such that $\alpha(u) = 1$ and $\alpha^*(u^{-1}) = N$. We claim that this implies

$$a_k(u) \geq 1 - k/N \text{ for } k = 1, 2, \ldots, N.$$

Indeed, let P_1 be any orthogonal projection on ℓ_2^N with rank $P < k$. Then

$$
\begin{aligned}
N - k < \operatorname{rank}(I - P) &= tr(I - P) \\
&= tr\ u^{-1}u(I - P) \\
&\leq \alpha^*(u^{-1})\|u(I - P)\| \\
&\leq N\|u - uP\|.
\end{aligned}
$$

Therefore we have $\|u - uP\| > 1 - k/N$, hence $a_k(u) \geq 1 - k/N$. By Lemma 1.8, there is an orthonormal basis $f_1, \ldots, f_N$ of ℓ_2^N such that $\|uf_k\| \geq a_k(u) \geq 1 - k/N$ for all k. Let $x_k = u(f_k)$. Then (since (f_k) is orthonormal and $\|u\| \leq 1$) we have

$$
\forall\, \alpha \in \mathbf{R}^N \quad \left\| \sum \alpha_k x_k \right\| \leq \left(\sum |\alpha_k|^2 \right)^{1/2}.
$$

A fortiori, (4.11) holds. On the other hand, if $k \leq [N/2]$ we have $\|x_k\| \geq 1/2$. ∎

Proof of Theorem 4.3: Let (g_k) be a sequence of i.i.d. standard Gaussian r.v.'s as before. Let then $(x_k)_{k \leq \overline{N}}$ be as in the preceding Lemma 4.13. Let

$$
X = \sum_{k \leq \overline{N}} g_k x_k.
$$

We claim

(4.13) $$ d(X) \geq c(\operatorname{Log} N) $$

for some numerical constant $c > 0$. Using (4.13) and applying Theorem 4.4, we immediately obtain Theorem 4.3.

It remains to prove (4.13). First note that (4.11) implies $\sigma(X) \leq 1$. On the other hand, we have the following elementary fact:

(4.14) $$ \mathbf{E} \sup_{k \leq \overline{N}} |g_k|\ \|x_k\| \leq \mathbf{E}\|X\|. $$

Indeed, let A_k be a partition of our probability space such that

$$
A_k \subset \{\omega|\ \ |g_k(\omega)|\|x_k\| = \sup_{i \leq \overline{N}} \|g_i(\omega)x_i\|\}.
$$

Then

$$
\sup_{i \leq \overline{N}} |g_i(\omega)|\ \|x_i\| = \left\| \sum_k g_k 1_{A_k} x_k \right\|,
$$

so that

$$\mathbf{E} \sup_{i \leq \overline{N}} |g_i(\omega)| \, \|x_i\| = \mathbf{E} \left\| \sum g_k 1_{A_k} x_k \right\|$$

$$\leq \mathbf{E} \left\| \sum g_k x_k \right\|$$

(the last line follows from the triangle inequality and the fact that, by symmetry, $\sum g_k (1 - 21_{A_k}) x_k$ has the same distribution as X).

This establishes (4.14). Now (4.12) implies

$$\frac{1}{2} \mathbf{E} \sup_{k \leq \overline{N}} |g_k| \leq \mathbf{E} \|X\|.$$

By an elementary computation, there is a constant $C' > 0$ such that

$$\mathbf{E} \sup_{i \leq n} |g_i| \geq C' (\text{Log } n)^{1/2} \text{ for all } n$$

(see Lemma 4.14 below). Therefore we conclude that (4.13) holds. ∎

The following elementary lemma will allow us to compute easily the number $\delta(E)$ when $E = \ell_q^N$.

Lemma 4.14. *Let $1 \leq q < \infty$. Let $\gamma(q) = (\mathbf{E}|g_1|^q)^{1/q}$ as in Chapter 2. Let $Z_1, \ldots, Z_N$ be N real-valued Gaussian variables (with mean zero but arbitrary variance). Then*

$$(4.15) \quad \mathbf{E} \left(\sum_1^N |Z_k|^q \right)^{1/q} \leq \left(\sum_1^N \mathbf{E}|Z_k|^q \right)^{1/q} \leq \gamma(q) N^{1/q} \sup_k \|Z_k\|_2$$

and

$$(4.16) \quad \mathbf{E} \sup_{k \leq N} |Z_k| \leq c(1 + \text{Log } N)^{1/2} \sup_k \|Z_k\|_2,$$

where $c > 0$ is a numerical constant. On the other hand, for independent standard variables $g_1, \ldots, g_N$ we have

$$(4.17) \quad \left(\frac{2}{\pi} \right)^{1/2} N^{1/q} \leq \mathbf{E} \left(\sum_1^N |g_i|^q \right)^{1/q}$$

and

$$(4.18) \quad c'(\text{Log } N)^{1/2} \leq \mathbf{E} \sup_{k \leq N} g_k \leq \mathbf{E} \sup_{k \leq N} |g_k|.$$

Proof: A Gaussian variable Z_k is a multiple of a standard Gaussian variable. Hence, the distribution of each Z_k is the same as that of $\|Z_k\|_2 g_1$, so that $\|Z_k\|_q = \|Z_k\|_2 \gamma(q)$ for each k. Let

$$Y = \left(\sum_1^N |Z_k|^q \right)^{1/q}.$$

Using $\|Y\|_1 \leq \|Y\|_q$, (4.15) is immediate. For some constant K we have $\gamma(q) \leq K q^{1/2}$ for all $q \geq 1$. Therefore (4.15) implies

$$\mathbf{E} \sup_{k \leq N} |Z_k| \leq K q^{1/2} N^{1/q} \sup_{k \leq N} \|Z_k\|_2.$$

Hence (4.16) follows from the choice $q = 1 + \mathrm{Log}\ N$. The inequality (4.17) is easy and left to the reader (recall $\gamma(1) = (2/\pi)^{1/2}$).

In view of the tail behaviour of Gaussian variables, it is easy to find a positive number $\alpha > 0$ such that

$$\mathbf{P}\{g_1 > \alpha(\mathrm{Log}\ N)^{1/2}\} \geq 1/N \text{ for all } N > 1.$$

Hence

$$\mathbf{P} \left\{ \sup_{k \leq N} g_k \leq \alpha(\mathrm{Log}\ N)^{1/2} \right\} = \left(\mathbf{P}\{g_1 \leq \alpha(\mathrm{Log}\ N)^{1/2}\} \right)^N$$

$$\leq (1 - 1/N)^N \leq e^{-1}$$

from which it is easy to derive (4.18). ■

We now estimate $\delta(E)$, or, equivalently, $n_\varepsilon(E)$ for $E = \ell_q^N$.

Theorem 4.15.

(i) *There are functions $\alpha(\varepsilon) > 0$ and $\beta(\varepsilon) > 0$ such that for all $N > 1$ we have*

(4.19) $\alpha(\varepsilon) \mathrm{Log}\ N \leq n_\varepsilon(\ell_\infty^N) \leq \beta(\varepsilon) \mathrm{Log}\ N.$

(ii) *For each $2 \leq q < \infty$, there are functions $\alpha(q, \varepsilon) > 0$ and $\beta(q, \varepsilon) > 0$ such that for all $N > 1$*

$$\alpha(q, \varepsilon) N^{2/q} \leq n_\varepsilon(\ell_q^N) \leq \beta(q, \varepsilon) N^{2/q}.$$

(iii) *For each $1 \leq q \leq 2$, there is a function $\phi(\varepsilon)$ such that for all $N > 1$*

$$\phi(\varepsilon) N \leq n_\varepsilon(\ell_q^N) \leq N.$$

Proof: Fix q such that $1 \leq q \leq \infty$. Let (e_k) be the canonical basis of ℓ_q^N and let $X = \sum_1^N g_k e_k$. If we consider X as a random variable with values in ℓ_q^N, then we have, by Lemma 4.14,

$$d(X) \sim \begin{cases} \operatorname{Log} N, & \text{for } q = \infty; \\ N^{2/q}, & \text{for } 2 \leq q < \infty; \\ N, & \text{for } 1 \leq q \leq 2. \end{cases}$$

(Indeed, note that $\sigma(X) \leq 1$ for $2 \leq q \leq \infty$ and $\sigma(X) \leq N^{1/q-1/2}$ for $1 \leq q \leq 2$.)

Thus the lower bounds of Theorem 4.15 all follow from Theorem 4.4. For the upper bounds, we use Proposition 4.6. We note that Lemma 4.14 implies

$$\delta(\ell_\infty^N) \leq C \, \operatorname{Log} N \quad \text{and} \quad \delta(\ell_q^N) \leq \gamma(q)^2 N^{2/q}.$$

By Proposition 4.6, this yields the first two upper bounds. The third one is trivial. ∎

Remarks:

(i) The first part of the above theorem shows that the Log N estimate in Dvoretzky's Theorem is sharp: We cannot replace in Theorem 4.3 Log N by a function of N which is $o(\operatorname{Log} N)$ when $N \to \infty$.

(ii) However, the second part of the preceding theorem suggests that the Log N estimate can be improved for spaces which are *far* from ℓ_∞^N-spaces. This is indeed the case; see Theorem 10.14 in Chapter 10.

(iii) A careful examination of the above proof shows that it yields Theorem 4.3 with $\eta(\varepsilon) \sim \varepsilon^2 (\operatorname{Log} 1/\varepsilon)^{-1}$ when $\varepsilon \to 0$. (Indeed, $\delta(\varepsilon) \sim \varepsilon$ and we may use

$$(1 + 2/\delta)^n \leq \exp(n \, \operatorname{Log}(1 + 2/\delta))$$

instead of the rougher majorization in the proof.) Recently, by a different proof based on a new variant of Slepian's Lemma (which is Lemma 5.7 in the next chapter), Y. Gordon was able to obtain this result with $\eta(\varepsilon) \sim \varepsilon^2$ which is the optimal behaviour when $\varepsilon \to 0$; cf. [Go1].

(iv) Let F Be an arbitrary n-dimensional normed space. Let $(\zeta_i)_{i \le m}$ be a δ-net in the unit sphere of F^* with $0 < \delta < 1$ and $m \le (1 + 2/\delta)^n$ (cf. Lemma 4.10). Then we have

$$(4.20) \qquad \forall\, x \in F \quad (1 - \delta)\|x\| \le \sup_{i \le m} |\zeta_i(x)| \le \|x\|.$$

Indeed, for any x in F, there is a ζ in F^* with $\|\zeta\| = 1$ such that $\|x\| = \zeta(x)$. Choose ζ_i such that $\|\zeta - \zeta_i\| < \delta$. Then

$$\begin{aligned} \|x\| &= \zeta_i(x) + (\zeta - \zeta_i)(x) \\ &\le \sup_{i \le m} |\zeta_i(x)| + \delta\|x\| \ , \end{aligned}$$

which yields (4.20) immediately.

This shows that for *any* n-dimensional normed space F, there is a subspace $\widetilde{F}$ of ℓ_∞^m with $m \le (1 + 2/\delta)^n$ such that $d(F, \widetilde{F}) \le (1 - \delta)^{-1}$. In particular (choose $F = \ell_2^n, 1 + \varepsilon = (1 - \delta)^{-1}$ and $N = [(1 + 2/\delta)^n]$), this elementary argument shows that

$$n_\varepsilon(\ell_\infty^N) \ge c(\mathrm{Log}\ N)(\mathrm{Log}\,(1 + 2/\varepsilon))^{-1}$$

for some numerical constant $c > 0$. This is an alternate proof of the left side of (4.19). Note, however, that this uses a very special property of ℓ_∞^N and gives nothing concerning Theorem 4.3 for a general normed space E.

In the sequel we will use repeatedly the following refinement of Lemma 4.10.

Lemma 4.16. *Let* $\| \ \|_1$ *and* $\| \ \|_2$ *be two arbitrary norms on* $\mathbf{R}^n$ *with respective unit balls* B_1 *and* B_2. *Let* $\delta > 0$. *Then there is a finite set* $A \subset B_1$ *with*

$$(4.21), \qquad \mathrm{card}\ (A) \le \frac{\mathrm{vol}((2/\delta)B_1 + B_2)}{\mathrm{vol}(B_2)}$$

which is a δ-net *for* B_1 *with respect to* $\| \ \|_2$, *that is to say* $\forall\, x \in B_1 \ \exists\, y \in A$ *such that* $\|x - y\|_2 < \delta$. *In particular, if* $B_2 \subset B_1$ *we find*

$$(4.22) \qquad \mathrm{card}\ (A) \le (1 + 2/\delta)^n \frac{\mathrm{vol}(B_1)}{\mathrm{vol}(B_2)}.$$

Conversely, if A is any δ-net for B_1 with repect to $\| \ \|_2$ we have necessarily

(4.23) $$\delta^{-n} \frac{\mathrm{vol}(B_1)}{\mathrm{vol}(B_2)} \leq \mathrm{card}\ (A).$$

Proof: Let $(y_i)_{i \leq m}$ be a maximal subset of B_1 such that $\|y_i - y_j\|_2 \geq \delta$ for all $i \neq j$. Clearly, by maximality $A = \{y_i | i \leq m\}$ is a δ-net of B_1 for $\| \ \|_2$. The balls $y_i + 2^{-1}\delta B_2$ are disjoint and all contained in $B_1 + 2^{-1}\delta B_2$. Therefore, taking volumes, we obtain

$$\sum_{i=1}^{m} \mathrm{vol}(y_i + 2^{-1}\delta B_2) \leq \mathrm{vol}(B_1 + 2^{-1}\delta B_2);$$

hence

$$m(2^{-1}\delta)^n\ \mathrm{vol}(B_2) \leq \mathrm{vol}(B_1 + 2^{-1}\delta B_2),$$

and this implies (4.21).

In the particular case $B_2 \subset B_1$, we have

$$\frac{2}{\delta}B_1 + B_2 \subset B_1(\frac{2}{\delta} + 1);$$

hence $\mathrm{vol}((2/\delta)B_1 + B_2) \leq \mathrm{vol}(B_1)(2/\delta + 1)^n$ and (4.22) follows.

Finally, if

$$B_1 \subset \bigcup_{y \in A} \{y + \delta B_2\}$$

then necessarily $\mathrm{vol}(B_1) \leq \mathrm{card}(A)\delta^n\ \mathrm{vol}(B_2)$ and (4.23) follows immediately. ∎

Notes and Remarks

The original proof of Dvoretzky's Theorem (Theorem 4.1 and Corollary 4.2) first appeared in [Dv]. Later, several simplified proofs appeared, namely in [M6], [Sz], and [F]. Milman's proof was the first one to give the Log N estimate of Theorem 4.3 (which is sharp), while the original proof only gave (Log $N)^{1/2}$. Milman's approach was considerably broadened in the paper [FLM], which had an enormous influence on the developments of the *Local Theory* of Banach Spaces during the past decade. Both [M6] and [FLM] make crucial use of Paul Lévy's isoperimetric inequality on the Euclidean sphere. In Milman's terminology, this inequality gives rise to a *concentration of measure phenomenon*: the mass on the Euclidean sphere is strongly concentrated near an equator. Stated analytically (as a Sobolev type inequality), this says that a Lipschitz function $f : S_{n-1} \to \mathbf{R}$ with $\|\operatorname{grad} f\|_\infty \leq 1$ deviates from its median (or its mean) only on a set of relatively small measure for the normalized surface measure on S_{n-1}. (See Remark 4.8.)

Let us denote by m_n the normalized surface measure on a Euclidean sphere *of radius* $\sqrt{n}$ in $\mathbf{R}^n$.

It is known that m_n is closely related to the canonical Gaussian measure on $\mathbf{R}^n$ which we denote by γ_n. Precisely, let us denote by $\pi_k : \mathbf{R}^n \to \mathbf{R}^k$ the projection onto the k first coordinates. Then, by a classical observation of Poincaré, for each fixed k, the measures $\pi_k(m_n)$ tend weakly to γ_k when $n \to \infty$. Thus, provided one makes the right normalization, it is possible to deduce from Paul Lévy's inequality a deviation inequality for the Gaussian measures γ_k. This approach was developed by C. Borell, who obtained several beautiful applications of this simple idea cf. [Bo]. The Gaussian version of Paul Lévy's inequality is sometimes called Borell's inequality in the literature on Gaussian processes, where it has proved to be a very powerful tool.

It should be emphasized that Lévy's isoperimetric inequality on S_{n-1} is not so easy to prove as the classical case of $\mathbf{R}^n$. The proof which appears in the appendix of [FLM] is rather short, but it uses not so easy symmetrization techniques (analogous to the Schwarz or Steiner symmetrizations). Using a suitable transposition of these techniques to the Gaussian case, A. Ehrhard [Eh] obtained a direct proof of the Gaussian isoperimetric inequality of Borell without using the Poincaré limiting argument. Again, however, his proof is not so simple. More recently, B. Maurey and the author found a very simple proof of the deviation inequality which is required to prove Dvoretzky's Theorem using Gaussian variables. This is included above as Theorem 4.7. We refer to [P1] for more variations on the same theme, in particular more general Sobolev type inequalities for Gaussian measures are given in [P1]. Theorem 4.4 was already formulated in [P1], but it is nothing but a Gaussian reformulation of the corresponding result from [FLM], modulo the replacement of Lévy's inequality by Theorem 4.7. We have briefly described the original approach of [M6] or [FLM] (which uses the measure σ_m instead of γ_m) in Remark 4.8. I have been informed that Proposition 4.6 was observed at an early stage by Figiel (unpublished).

Corollary 4.9 is a by now classical result due independently to Fernique [Fe2] and Landau–Shepp [LS]. We note in passing that [LS] already used the isoperimetric inequality on the Euclidean sphere to prove this. Lemmas 4.10 and 4.11 and Remark 4.12 are elementary classical facts as well as Lemma 4.16. Similarly, Lemma 4.14 contains only elementary inequalities in Probability Theory. Lemma 4.13 is a variant of the classical Dvoretzky–Rogers Lemma [DvR]. Similar variants appear in [FLM]. We have borrowed Lewis's ideas in [L] for this simple proof. We refer the interested reader to [To] for more information on inequalities involving Gaussian random vectors. (See also [MPi].) For more details on the isoperimetric inequalities see [GM2], [GM3] and the appendix of [MS] written by Gromov.

Chapter 5

Entropy, Approximation Numbers, and Gaussian Processes

Let $u : X \to Y$ be an operator between Banach spaces. Let $k \geq 1$ be an integer. Recall the definition of the approximation numbers of u

$$a_k(u) = \inf\{\|u - v\| \mid v : X \to Y \quad \mathrm{rk}(v) < k\}.$$

The *Gelfand numbers* of u are defined as

$$c_k(u) = \inf\left\{\|u_{|S}\| \mid S \subset X, \text{ codim } S < k, \right\}.$$

For any (closed) subspace $S \subset Y$ we denote by Q_S the quotient mapping from Y onto Y/S. We define the *Kolmogorov numbers* of u as

$$d_k(u) = \inf\{\|Q_S u\| \mid S \subset Y \quad \dim S < k\}.$$

The sequences $\{a_k(u)\}, \{c_k(u)\}, \{d_k(u)\}$ are all non-increasing and satisfy $a_1(u) = c_1(u) = d_1(u) = \|u\|, c_k(u) \leq a_k(u), d_k(u) \leq a_k(u)$. Moreover, we have $c_k(u) = d_k(u^*)$ and if u is compact, we have $d_k(u) = c_k(u^*)$ and $a_k(u) = a_k(u^*)$.

Let s_k denote either a_k, d_k or c_k. It is easy to check that for all operators $u : X \to Y$ and $v : Y \to Z$ between Banach spaces we have

$$(5.1) \qquad s_{k+n-1}(vu) \leq s_k(v)s_n(u) \quad \forall \, k, n \geq 1.$$

Similarly, if u_1 and u_2 are operators from X into Y

$$(5.2) \qquad s_{k+n-1}(u_1 + u_2) \leq s_k(u_1) + s_n(u_2) \quad \forall \, k, n \geq 1.$$

If u is a compact operator, then $c_k(u) \to 0$ and $d_k(u) \to 0$ when $k \to \infty$ and the speed of this convergence provides a *measure* of the degree of compactness of u. We will also need other numbers called the *entropy numbers* which we now define.

Let K_1, K_2 be two subsets of a Banach space Y. We will denote by $N(K_1, K_2)$ the smallest number N such that there are points $y_1, \ldots, y_N$ in Y such that $K_1 \subset \cup_{i \leq N}(y_i + K_2)$.

Let B_X denote the closed unit ball of X.

We define the entropy numbers of $u : X \to Y$ as follows:

$$e_k(u) = \inf \left\{ \varepsilon > 0 \big| N(u(B_X), \varepsilon B_Y) \leq 2^{k-1} \right\}.$$

Note that $\|u\| = e_1(u) \geq e_2(u) \geq \ldots$. Clearly u is compact iff $e_k(u) \to 0$ when $k \to \infty$. We note the following elementary fact valid when K_1, K_2, K_3 are arbitrary subsets of a space Y:

(5.3) $$N(K_1, K_3) \leq N(K_1, K_2)N(K_2, K_3).$$

This fact immediately implies that for a composition of two operators $u : X \to Y$ $v : Y \to Z$ between Banach spaces, we have

(5.4) $$e_{n+k-1}(vu) \leq e_n(u)e_k(v).$$

Similarly, we have obviously, if K_4 is another subset of Y,

(5.5) $$N(K_1 + K_2, K_3 + K_4) \leq N(K_1, K_3)N(K_2, K_4),$$

and this implies for all operators u_1, u_2 from X into Y

(5.6) $$e_{n+k-1}(u_1 + u_2) \leq e_n(u_1) + e_k(u_2).$$

Thus the entropy numbers behave very much like the numbers a_k, c_k and d_k. Let s_k denote either a_k, c_k, d_k, or e_k.

Let $S_p(X, Y)$ be the class of all operators u from X into Y such that $\sum s_k(u)^p < \infty$. Then an easy appliction of the preceding inequalities yields, with the same notation as above,

(i) $$u \in S_p(X, Y), v \in S_q(Y, Z) \Rightarrow vu \in S_r(X, Z)$$

with
$$\frac{1}{r} = \frac{1}{p} + \frac{1}{q}, \quad p > 0, q > 0, r > 0.$$

(ii) $$u_1, u_2 \in S_p(X, Y) \Rightarrow u_1 + u_2 \in S_p(X, Y).$$

Moreover, we have $s_k(vu) \leq \|v\|s_k(u)$ and $s_k(vu) \leq \|u\|s_k(u)$ so that $S_p(X, Y)$ has the *ideal property*, i.e. it is stable by two-sided composition with bounded operators.

Let E be an n-dimensional normed space, then clearly $c_k(u) = d_k(u) = a_k(u) = 0$ for all $k > n$ and all operators u on E. For the entropy numbers the same fact is clearly wrong, but an analogous phenomenon holds: the numbers $e_k(u)$ decrease *very fast* when $k > n$. This is best seen by studying the identity operator I_E. Let B_E be the unit ball of E. We note that for all $0 < \varepsilon \leq 1$,

$$(5.7) \qquad N(B_E, \varepsilon B_E) \leq \left(1 + \frac{2}{\varepsilon}\right)^n .$$

Indeed, this follows from Lemma 4.16.

Clearly, (5.7) implies

$$(5.8)' \qquad e_k(I_E) \leq 2 \left(2^{\frac{k-1}{n}} - 1\right)^{-1} \text{ for all } k > 1.$$

On the other hand, (4.23) implies, for all $0 < \varepsilon \leq 1$,

$$\varepsilon^{-n} \leq N(B_E, \varepsilon B_E),$$

and therefore

$$(5.8)'' \qquad\qquad 2^{-(k-1)/n} \leq e_k(I_E)$$

for all $k \geq 1$.

We refer to [Pi1], [Pi3], or [Kö] for more details on the numbers $(s_k(u))$.

We will need the following elementary fact:

Proposition 5.1.   *Let X, X_1, Y, Y_1 be Banach spaces. Assume that X is isometric to a quotient of X_1, and that Y is isometric to a subspace of Y_1. We denote by $q : X_1 \to X$ the quotient map and by $j : Y \to Y_1$ the isometric embedding. Then for any operator $u : X \to Y$ we have*

$$e_k(u) = e_k(uq)$$

and

$$\frac{1}{2} e_k(u) \leq e_k(ju) \leq e_k(u) \text{ for all } k \geq 1.$$

Moreover, if $X_1 = \ell_1(I)$ for some set I we have $d_k(u) = a_k(uq)$, while if $Y_1 = \ell_\infty(I)$ we have $a_k(ju) = c_k(u)$.

Proof: The first part is immediate since $q(B_{X_1}) = B_X$. Also, $e_k(ju) \leq \|j\| e_k(u) = e_k(u)$. Note that if $ju(B_X)$ is covered by 2^{k-1} balls of radius ε in Y_1, then $u(B_X)$ is the union of 2^{k-1} subsets of diameter

2ε, hence $e_k(u) \leq 2\, e_k(ju)$. Finally, the last assertions are immediate consequences of the lifting property of $\ell_1(I)$ and the extension property of $\ell_\infty(I)$. ∎

The following result of Carl will be very useful in the sequel. It shows that the numbers c_k, d_k, or a_k *dominate* in a certain sense the entropy numbers.

Theorem 5.2. *Let s_k denote either c_k, d_k, or a_k. For each $\alpha > 0$ there is a constant ρ_α such that for all operators $u : X \to Y$ between Banach spaces we have*

$$(5.9) \qquad \sup_{k \leq n} k^\alpha e_k(u) \leq \rho_\alpha \sup_{k \leq n} k^\alpha s_k(u) \ \text{ for all } \ n \geq 1.$$

More generally, for any $0 < q < \infty$ there is a constant $\rho_{\alpha q}$ such that for all u as above we have

$$(5.10) \qquad \left(\sum k^{-1} (k^\alpha e_k(u))^q \right)^{1/q} \leq \rho_{\alpha q} \left(\sum k^{-1} (k^\alpha s_k(u))^q \right)^{1/q}.$$

In particular, there is a constant ρ_q' such that

$$\sum e_k(u)^q \leq \rho_q' \sum s_k(u)^q.$$

Proof: By the preceding proposition, it is enough to prove this lemma for $s_k = a_k$. (Recall that any Banach space X is isometric to a quotient of $\ell_1(I)$, and similarly any Y embeds isometrically into $\ell_\infty(I)$ for some set I).

Consider $u : X \to Y$. We assume $\sup_{k \leq n} k^\alpha a_k(u) < 1$ and $n = 2^N$ for some $N \geq 0$. For any m there is an operator $v_m : X \to Y$ with rank $< 2^m$ such that $\|u - v_m\| < 2^{-m\alpha}$.

Let $\Delta_0 = v_0, \ldots, \Delta_m = v_m - v_{m-1}$, etc., so that we have $u = \sum_{m \leq N} \Delta_m + u - v_N$ and we have $\|\Delta_m\| \leq 2^{\alpha+1} 2^{-m\alpha}$ and $\mathrm{rk}(\Delta_m) < 2^{m+1}$. Let $K = u(B_X), K_m = \Delta_m(B_X)$ and let $\varepsilon_m > 0$ to be specified later. We deduce from (5.7) since $K_m \subset \|\Delta_m\| B_Y$, that

(5.11)

$$\forall \varepsilon > 0 \quad N(K_m, \varepsilon \|\Delta_m\| B_Y) \leq \left(1 + \frac{2}{\varepsilon} \right)^{\mathrm{rk}(\Delta_m)} \leq \left(1 + \frac{2}{\varepsilon} \right)^{2^{m+1}}.$$

Let $0 < r < 1$ to be specified later, and note that (since $(1+t)^d \leq (1+t^r)^{d/r}$)

$$\left(1 + \frac{2}{\varepsilon}\right)^{2^{m+1}} < \exp\left(\left(\frac{2}{\varepsilon}\right)^r \frac{2^{m+1}}{r}\right).$$

By the triangle inequality we have

$$K \subset \sum_{m=0}^{N} K_m + (u - v_N)B_X.$$

For simplicity, let $N(\varepsilon) = N(K, \varepsilon B_Y)$. Since $\|u - v_N\| < 2^{-N\alpha}$, by (5.5) and (5.11) we have

(5.12)
$$N\left(\sum_{m=1}^{N} \varepsilon_m \|\Delta_m\| + 2^{-N\alpha}\right) \leq \prod_{m=0}^{N} N\left(K_m, \varepsilon_m \|\Delta_m\| B_Y\right)$$

$$< \exp\left\{\sum_{m=0}^{N} \left(\frac{2}{\varepsilon_m}\right)^r \frac{2^{m+1}}{r}\right\}.$$

Now choose $\beta > \alpha$ and $0 < r < 1$ small enough so that $r < 1/\beta$. Let $t > 0$ be arbitrary. We make the choice $\varepsilon_m = t2^{m\beta}2^{-N\beta}$. This gives

$$\sum_{0}^{N} \varepsilon_m \|\Delta_m\| + 2^{-N\alpha} \leq 2^{-N\alpha}(c_1 t + 1)$$

and

$$\sum_{m=0}^{N} \left(\frac{2}{\varepsilon_m}\right)^r \frac{2^{m+1}}{r} \leq c_2 t^{-r} 2^N$$

for some constants c_1, c_2, depending only on α, β and r.

Finally, we choose $t = T$, with T large enough (T depends on α, β, r) so that

$$2\exp(c_2 t^{-r} 2^N) \leq 2^{2^N} = 2^n.$$

Then (5.12) implies

$$e_n(u) \leq 2^{-N\alpha}(c_1 T + 1) = n^{-\alpha}(c_1 T + 1).$$

By homogeneity, we have proved for $n = 2^N$

$$n^\alpha e_n(u) \leq (c_1 T + 1) \sup_{k \leq n} k^\alpha a_k(u);$$

clearly (5.9) follows since it is easy to extend the last result to arbitrary values of n.

We now turn to the second estimate (5.10). We note that $\sum k^{-1}(k^\alpha a_k(u))^q$ is clearly equivalent to $\sum (2^{n\alpha} a_{2^n}(u))^q$, and similarly for $(e_k(u))$.

Assume that

$$\sum k^{-1}(k^\alpha a_k(u))^q \leq 1.$$

Then

$$\sum_{j \geq 0} \left(2^{j\alpha} a_{2^j}(u)\right)^q \leq c_3$$

for some constant c_3 depending only on α and q.

Choose $\beta > \alpha$. Let

$$\lambda_n = 2^{-n\beta} \sup_{0 \leq k \leq n} \left(2^{k\beta} a_{2^k}(u)\right).$$

Using the obvious inequality,

$$\lambda_n^q \leq 2^{-n\beta q} \sum_{k \leq n} 2^{k\beta q} a_{2^k}(u)^q,$$

we find

$$\sum_{n \geq 0} (2^{n\alpha}\lambda_n)^q \leq \sum_{n \geq 0} 2^{n\alpha q} \left[2^{-n\beta q} \sum_{k \leq n} 2^{k\beta q} a_{2^k}(u)^q\right]$$

$$\leq c_4 \sum_{k \geq 0} 2^{k\alpha q} a_{2^k}(u)^q \leq c_4 c_3.$$

By the first part of the proof, we know that

$$e_{2^n}(u) \leq \rho_\beta'' \lambda_n$$

for some constant ρ_β''. Thus we conclude that

$$\left(\sum (2^{n\alpha} e_{2^n}(u))^q\right)^{1/q} \leq c_4 c_3 \rho_\beta'',$$

so that finally

$$\left(\sum k^{-1}(k^\alpha e_k(u))^q\right)^{1/q} \leq c_5.$$

By homogeneity, this concludes the proof of (5.10).

The last assertion is a particular case of (5.10) (just take $\alpha q = 1$). ∎

Remark: Note that (5.10) compares the quasi-norms of $(e_k(u))$ and $(s_k(u))$ in the Lorentz sequence space ℓ_{pq} for $p = \alpha^{-1}$.

Let $T : X \to X$ be a compact operator and let $\lambda_n(T)$ be the eigenvalues of T rearranged as usual so that $|\lambda_n(T)|$ is non-increasing and each eigenvalue is repeated according to its multiplicity.

Let us consider first the special case when X is of dimension n over $\mathbf{R}$ and $T : X \to X$ is $\mathbf{R}$-linear with (possibly complex) eigenvalues $\lambda_1(T), \ldots, \lambda_n(T)$.

Then we clearly have

$$\frac{\text{vol}(T(B_X))}{\text{vol}(B_X)} = |\det(T)| = \Big|\prod_{i=1}^{n} \lambda_i(T)\Big|.$$

Here the volume is meant in $\mathbf{R}^n$.

Using the definition of $e_n(T)$, we find

$$\text{vol}(T(B_X)) \leq e_n(T)^n 2^{n-1} \text{vol}(B_X);$$

hence

$$|\lambda_n(T)| \leq \Big|\prod \lambda_i(T)\Big|^{1/n} \leq 2 e_n(T).$$

Now assume more generally that X is of dimension n over $\mathbf{C}$. Then, using the volume in $\mathbf{C}^n = \mathbf{R}^{2n}$ we find

$$\frac{\text{vol}(T(B_X))}{\text{vol}(B_X)} = \prod_{1}^{n} |\lambda_i(T)|^2;$$

hence the same argument gives us in this case

$$\prod_{i=1}^{N} |\lambda_i(T)|^2 \leq 2^{n-1} e_n(T)^{2n},$$

hence

$$|\lambda_n(T)| \leq \left(\prod_{i=1}^{n} |\lambda_i(T)|\right)^{1/n} \leq 2^{1/2} e_n(T).$$

In the sequel, when dealing with eigenvalues we always assume that we are in the complex case. By a simple variant we find.

Theorem 5.3. *Let $T : X \to X$ be a compact operator on a complex Banach space. Then for all $n \geq 1$*

$$|\lambda_n(T)| \leq \Big|\prod_{i=1}^{n} \lambda_i(T)\Big|^{1/n} \leq 2^{3/2} e_n(T).$$

Proof: Let E be a spectral subspace for T of dimension n such that $T_{|E} : E \to E$ admits exactly $\lambda_1(T), \ldots, \lambda_n(T)$ as its eigenvalues. Then Theorem 5.3 follows immediately from the preceding remark since $e_n(T_{|E}) \leq 2 e_n(T)$ by Proposition 5.1.

Corollary 5.4. *Let $T : H_1 \to H_2$ be a compact operator between two Hilbert spaces. Let $0 < \alpha < \infty$ and $0 < q < \infty$. Then*

$$\sum_{k \geq 1} k^{-1}(k^\alpha e_k(T))^q < \infty \text{ iff } \sum_{k \geq 1} k^{-1}(k^\alpha \lambda_k(|T|))^q < \infty.$$

In particular,

$$(e_k(T)) \in \ell_q \text{ iff } (\lambda_k(|T|)) \in \ell_q.$$

Proof: The *if* part follows from Theorem 5.2 (recall (1.12)). For the converse, we may use the polar decomposition $T = U|T|$ of T. By Theorem 5.3, we find

$$\lambda_n(|T|) \leq 4e_n(|T|) = 4e_n(U^*T)$$
$$\leq 4e_n(T) \text{ since } \|U^*\| \leq 1.$$

From this the converse implication follows immediately. ∎

Let $u : H \to X$ be an operator defined on a Hilbert space H. If $\dim H = n < \infty$, we may identify H with ℓ_2^n and we have already defined

$$\ell(u) = \left(\int_{\mathbf{R}^n} \|u(x)\|^2 \, d\gamma_n(x)\right)^{1/2}.$$

If H is infinite–dimensinal, we have set

$$\ell(u) = \sup\left\{\ell(u_{|_E}) \big| E \subset H \quad \dim E < \infty\right\}.$$

The next result gives a two-sided estimate for $\ell(u)$ in terms of entropy numbers.

Theorem 5.5. *There are absolute constants $C_1 > 0, C_2$ such that any operator $u : H \to X$ (from a Hilbert space H into a Banach space X) satisfies*

$$C_1 \sup_{n \geq 1} n^{1/2} e_n(u^*) \leq \ell(u) \leq C_2 \sum_{n \geq 1} n^{-1/2} e_n(u^*).$$

Remark: If $\ell(u)$ is finite, the preceding theorem implies $e_n(u) \to 0$ when $n \to \infty$. Hence in particular u is compact whenever $\ell(u)$ is finite.

Remark: Consider $u : \ell_2^n \to X$ and let $K = u^*(B_{X^*}) \subset \ell_2^n$. Then, denoting (e_k) the canonical basis of ℓ_2^n, we have clearly

$$\ell(u) = \left(\mathbf{E} \sup_{t \in K} \left| < t, \sum_1^n g_k e_k > \right|^2\right)^{1/2}.$$

Since $Z_t = <t, \sum_1^n g_k e_k> = \sum g_k t_k$ is a Gaussian random variable, we see that $\ell(u)$ is nothing but the L_2-norm of the supremum of a family of Gaussian r.v.s., and Theorem 5.5 is nothing but a reformulation of a classical theorem in the theory of Gaussian stochastic processes.

Let us introduce some terminology.

Let $\{Z_i | i \in I\}$ be a collection of real-valued random variables indexed by a set I and defined on a probability space $(\Omega, \mathcal{A}, \mathbf{P})$. We say that $\{Z_i | i \in I\}$ is a Gaussian process if all the linear combinations of the variables $\{Z_i\}$ are Gaussian. (For instance (with the notation of the preceding remark) if we denote $Z_t = \sum_1^n g_k t_k$ for $t \in \mathbf{R}^n$ then $\{Z_t | t \in \mathbf{R}^n\}$ is a Gaussian process.)

We denote by K_Z the set $\{Z_i | i \in I\}$ considered as a subset of L_2 (here L_2 means $L_2(\Omega, \mathcal{A}, \mathbf{P})$). For any $\varepsilon > 0$, we denote simply

$$N_Z(\varepsilon) = N(K_Z, \varepsilon B_{L_2}).$$

Then we can state:

Theorem 5.6. *There are absolute constants $C_1' > 0$ and C_2' such that for any Gaussian process $Z = \{Z_i | i \in I\}$, indexed by a finite or countable set I, we have*
(5.13)

$$C_1' \sup_{\varepsilon > 0} \varepsilon \left(\text{Log } N_Z(\varepsilon)\right)^{1/2} \leq \mathbf{E} \sup_{i \in I} Z_i \leq C_2' \int_0^\infty \left(\text{Log } N_Z(\varepsilon)\right)^{1/2} d\varepsilon.$$

Remark: More generally, the preceding result still holds if we merely assume Z separable, i.e. such that there is a countable subset $J \subset I$ for which $\widetilde{K} = \{Z_i | i \in J\}$ is dense in K_Z. Indeed, we have then $N(\widetilde{K}, \varepsilon B_{L_2}) = N_Z(\varepsilon)$. If we define $\sup_{i \in I} Z_i$ as the supremum in the Banach *lattice* L_1 then clearly $\sup_{i \in I} Z_i = \sup_{i \in J} Z_i$ a.s., and (5.13) still holds.

For the lower bound in (5.13), the following lemma is crucial. It originates essentially in the work of Slepian.

Lemma 5.7. *Let $\{X_i, 1 \leq i \leq n\}$ and $\{Y_i, 1 \leq i \leq n\}$ be two Gaussian processes such that*

(5.14) $$\forall\, i, j \quad \|Y_i - Y_j\|_2 \leq \|X_i - X_j\|_2,$$

then

$$\mathbf{E} \sup Y_i \leq 2\mathbf{E} \sup X_i.$$

If we assume moreover that

(5.15) $\forall\, i \le n \quad \|Y_i\|_2 = \|X_i\|_2,$

then for all $(c_i) \in \mathbf{R}^n$ *we have*
(5.16)
$$\mathbf{P}(\{Y_1 > c_1\} \cup \ldots \cup \{Y_n > c_n\}) \le \mathbf{P}(\{X_1 > c_1\} \cup \ldots \cup \{x_n > c_n\}).$$

Remark: The above result is known without the factor 2 but we do not use this improvement in the sequel. We refer the reader to [BC] and [Fe1] or [Kah1] and [Kah2] for more information.

Remark: It is worthwhile to observe by symmetry that

(5.17) $\mathbf{E} \sup_{i,j} |X_i - X_j| = \mathbf{E} \sup_{ij}(X_i - X_j) = 2\mathbf{E} \sup X_i.$

Proof of Lemma 5.7: We assume (5.14) and (5.15). We will first prove (5.16). We may clearly assume without loss of generality that $X = (X_i)_{i \le n}$ and $Y = (Y_i)_{i \le n}$ are independent of each other. We then introduce for any $0 \le t \le 1$ the random vector $X(t) = (X_i(t))_{i \le n}$ defined by
$$X_i(t) = (1 - t)^{1/2} X_i + t^{1/2} Y_i.$$
We set $p(t) = \mathbf{P}(X_1(t) \le c_1, \ldots, X_n(t) \le c_n).$

We will show that $p(t)$ is a non-decreasing function of t. We may clearly assume (by a perturbation argument) that the distribution of X is absolutely continuous, hence the same is true for $X(t)$. We will denote by φ_X the density of the Gaussian vector X.

Then the key to the proof lies in the following identity:
(5.18)
$$\forall\, y \in \mathbf{R}^n \quad \frac{d}{dt}\varphi_{X(t)}(y) = \sum_{i<j}(\mathbf{E}Y_iY_j - \mathbf{E}X_iX_j)\frac{\partial^2}{\partial y_i\,\partial y_j}\varphi_{X(t)}(y).$$

This identity is easy to check using the inverse Fourier transform formula

(5.19) $\varphi_{X(t)}(y) = \int \exp\{i < \zeta, y > -\frac{1}{2}\mathbf{E} < \zeta, X(t) >^2\}\, d\zeta,$

where $d\zeta$ denotes a (suitably normalized) Haar measure on $\mathbf{R}^n$.

Indeed, to check that (5.19) implies (5.18) all we need is to observe that (5.15) implies
$$\mathbf{E} < \zeta, X(t) >^2 = \mathbf{E}\Big|\sum \zeta_i X_i\Big|^2 + 2t \sum_{i<j} \zeta_i\zeta_j \mathbf{E}(Y_iY_j - X_iX_j).$$

Let $\gamma_{ij} = \mathbf{E} Y_i Y_j - \mathbf{E} X_i X_j$. Note that $\gamma_{ij} \geq 0$ by (5.13) and (5.15). Using (5.18), we find

$$\frac{d}{dt} p(t) = \frac{d}{dt} \int_{-\infty}^{c_1} \cdots \int_{-\infty}^{c_n} \varphi_{X(t)}(y) \, dy$$

$$= \int_{-\infty}^{c_1} \cdots \int_{-\infty}^{c_n} \sum_{i<j} \gamma_{ij} \frac{\partial^2 \varphi_{X(t)}}{\partial y_i \, \partial y_j}(y) \, dy.$$

But each term in the last sum is ≥ 0; indeed, for instance the term corresponding to $i = 1, j = 2$ is equal to

$$\gamma_{12} \int_{-\infty}^{c_3} \cdots \int_{-\infty}^{c_n} \varphi_{X(t)}(c_1, c_2, y_3, \ldots, y_n) \, dy_3 \ldots dy_n \geq 0;$$

it is positive since the density is positive. Therefore, we conclude that $p(0) \leq p(1)$ and this yields (5.16).

We now turn to the first part of the lemma. We may clearly assume (replacing Y_i by $Y_i - Y_1$ and X_i by $X_i - X_1$) that $Y_1 = 0$ and $X_1 = 0$ and $\mathbf{E} Y_i^2 \leq \mathbf{E} X_i^2$ for all $i \leq n$. Then let g be an auxiliary standard Gaussian variable independent of (X_i) and (Y_i). We set

$$C^2 = \sup_{i \leq n} \mathbf{E} X_i^2$$

and let

$$\widetilde{X}_i = X_i + g \left\{ C^2 - \mathbf{E} X_i^2 + \mathbf{E} Y_i^2 \right\}^{1/2}$$

(note the positivity of the coefficient of g) and

$$\widetilde{Y}_i = Y_i + gC.$$

We clearly have

$$\|\widetilde{Y}_i - \widetilde{Y}_j\|_2 = \|Y_i - Y_j\|_2 \leq \|X_i - X_j\|_2 \leq \|\widetilde{X}_i - \widetilde{X}_j\|_2$$

and

$$\|\widetilde{Y}_i\|_2 = \|\widetilde{X}_i\|_2.$$

Therefore, by the first part of the proof we have

$$\forall \, c \in \mathbf{R} \quad \mathbf{P}(\sup \widetilde{Y}_i > c) \leq \mathbf{P}(\sup \widetilde{X}_i > c);$$

integrating in c and using the following formula valid for all M in L_1:

$$\mathbf{E} M = \int_0^\infty \mathbf{P}(M > c) \, dc + \int_{-\infty}^0 [\mathbf{P}(M > c) - 1] \, dc,$$

we find

(5.20) $\mathbf{E} \sup \widetilde{Y}_i \leq \mathbf{E} \sup \widetilde{X}_i.$

Finally, we have

$$\mathbf{E} \sup \widetilde{Y}_i \leq \mathbf{E}(\sup Y_i + gC) = \mathbf{E} \sup Y_i$$

(5.21)   and

$$\mathbf{E} \sup \widetilde{X}_i \leq \mathbf{E} \sup X_i + \mathbf{E}g^+ C,$$

hence we obtain the announced result since

$$C = \sup_{i \leq n} \|X_i\|_2 = \sup_{i \leq n} \frac{\mathbf{E}X_i^+}{\mathbf{E}g^+} \leq (\mathbf{E}g^+)^{-1} \mathbf{E} \sup X_i^+$$

hence (recall $X_1 = 0$)

$$C \leq (\mathbf{E}g^+)^{-1} \mathbf{E} \sup X_i.$$

Therefore, (5.20) and (5.21) imply $\mathbf{E} \sup Y_i \leq 2\mathbf{E} \sup X_i.$ ∎

Proof of Theorem 5.6: It is clearly enough to prove the statement for I finite, say

$$I = \{1, 2, \dots, n\},$$

provided, of course, the constants we obtain do not depend on n.

We will use the following well-known fact (cf. Lemma 4.14): There is a constant $C > 0$ such that if (g_n) is a sequence of independent normal Gaussian variables we have

(5.22) $\forall n \geq 1 \quad C^{-1}(\text{Log } n)^{1/2} \leq \mathbf{E} \sup_{i \leq n} g_i \leq C(\text{Log } n)^{1/2}.$

This result implies that, if $(Z_1, \dots, Z_n)$ is an arbitrary finite Gaussian process, then

(5.23) $$(2\sqrt{2}C)^{-1}(\text{Log } n)^{1/2} \inf_{i \neq j \leq n} \|Z_i - Z_j\|_2 \leq$$
$$\mathbf{E} \sup_{i \leq n} Z_i \leq \sqrt{2}C \sup_{i \neq j \leq n} \|Z_i - Z_j\|_2 (\text{Log } n)^{1/2}.$$

Indeed, we may define

$$Y_i = g_i 2^{-1/2} \inf_{i \neq j} \|Z_i - Z_j\|_2$$

and
$$X_i = g_i 2^{-1/2} \sup_{i \neq j} \|Z_i - Z_j\|_2.$$

Then clearly,

$$\|Y_i - Y_j\|_2 \leq \|Z_i - Z_j\|_2 \leq \|X_i - X_j\|_2.$$

By Lemma 5.7,

$$\frac{1}{2} \mathbf{E} \sup Y_i \leq \mathbf{E} \sup Z_i \leq 2\mathbf{E} \sup X_i,$$

hence (5.23) follows from this and (5.22).

The left side of (5.13) follows immediately from (5.23). Indeed, let $\{Z_1, \ldots, Z_N\}$ be a maximal subset of $\{Z_i | i \in I\}$ such that $\|Z_i - Z_j\|_2 \geq \varepsilon$ for all $i \neq j \leq N$. By maximality, we have necessarily $N_Z(\varepsilon) \leq N$. On the other hand, by (5.23) we have

$$\varepsilon(2\sqrt{2}C)^{-1}(\mathrm{Log}\ N)^{1/2} \leq \mathbf{E} \sup_{i \leq N} Z_i \leq \mathbf{E} \sup_{i \in I} Z_i.$$

This proves the left side of (5.13) with $C_1' = (2\sqrt{2}C)^{-1}$.

Let us now turn to the right side. Let

$$D = \sup_{i,j \in I} \|Z_i - Z_j\|_2.$$

Let $\varepsilon_n = 2^{-n} D (n \geq 0)$ and let $N_n = N_Z(\varepsilon_n)$.

By definition of $N_Z(\varepsilon_n)$ we can find a subset $I_n \subset I$ with $\mathrm{card}(I_n) \leq N_n$ such that $\forall\ i \in I\ \exists\ j \in I_n$ such that $\|Z_i - Z_j\|_2 \leq 2\varepsilon_n$. Let us denote $j = \phi_n(i)$ in the preceding line and set $Z_i^n = Z_{\phi_n(i)}$.

We can write

$$\|Z_i - Z_i^n\|_2 \leq 2\varepsilon_n$$

and

$$Z_i = Z_i^\circ + \sum_{n \geq 1} Z_i^n - Z_i^{n-1};$$

note that Z_i° does not depend on i since $N_\circ = 1$. Let us denote $Z_i^\circ = z$. Hence

$$\sup_{i \in I} Z_i \leq z + \sum_{n \geq 1} \sup_{i \in I} \left(Z_i^n - Z_i^{n-1} \right),$$

so that

$$\mathbf{E} \sup_{i \in I} Z_i \leq \sum_{n \geq 1} \mathbf{E} \sup_{i \in I} \left(Z_i^N - Z_i^{n-1} \right).$$

But

$$\mathbf{E} \sup Z_i^n - Z_i^{n-1} \leq$$
$$\mathbf{E} \sup \{ Z_i - Z_j, i \in I_n, j \in I_{n-1}, \| Z_i - Z_j \|_2 \leq 4\varepsilon_n \},$$

hence by (5.17) and (5.23)

$$\leq 2\sqrt{2} C (\text{Log } N_n N_{n-1})^{1/2} 4\varepsilon_n \leq 16 C (\text{Log } N_n)^{1/2} \varepsilon_n.$$

This implies

$$\mathbf{E} \sup_{i \in I} Z_i \leq 16 C \sum_{n=1}^{\infty} \varepsilon_n (\text{Log } N_n)^{1/2}.$$

Comparing the last series with the integral

$$\int_0^{\infty} (\text{Log } N_Z(\varepsilon))^{1/2} \, d\varepsilon$$

we obtain the desired result. ∎

Proof of Theorem 5.5: It is easy to see that it suffices to prove these estimates for $u : \ell_2^n \to X$ with constants independent of n. For that purpose, we first note that by Corollary 4.9 we have

(5.24) $$\mathbf{E} \left\| \sum_1^n g_k u(e_k) \right\| \leq \ell(u) \leq K(2) \mathbf{E} \left\| \sum_1^n g_k u(e_k) \right\|.$$

Let $I = u^*(B_{X^*})$ and let $Z_\zeta = < \zeta, \sum_1^n g_k e_k >$ for ζ in I. Note that

$$\forall \, \zeta_1, \zeta_2 \in I \quad \| Z_{\zeta_1} - Z_{\zeta_2} \|_2 = \| \zeta_1 - \zeta_2 \|_{\ell_2^n},$$

so that

$$N_Z(\varepsilon) = N \left(u^*(B_{X^*}), \varepsilon B_{\ell_2^N} \right).$$

This implies that there are positive numerical constants α_1, α_2 such that
(5.25)
$$\alpha_1 \sum_{n=1}^{\infty} n^{-1/2} e_n(u^*) \leq \int_0^{\infty} (\text{Log } N_Z(\varepsilon))^{1/2} \, d\varepsilon \leq \alpha_2 \sum_{n=1}^{\infty} n^{-1/2} e_n(u^*)$$

and
(5.26)
$$\alpha_1 \sup n^{1/2} e_n(u^*) \leq \sup_{\varepsilon > 0} \varepsilon (\text{Log } N_Z(\varepsilon))^{1/2} \leq \alpha_2 \sup n^{1/2} e_n(u^*).$$

Finally, we note that

$$(5.27) \qquad \mathbf{E}\left\|\sum_1^n g_k u(e_k)\right\| = \mathbf{E}\sup_{\zeta \in I} Z_\zeta.$$

Then, recalling the remark following Theorem 5.6, it is easy to obtain Theorem 5.5 for $H = \ell_2^n$ by combining (5.13) with (5.24), (5.25), (5.26), and (5.27). Since the f.d. case suffices, the proof is complete. ∎

We will now improve Sudakov's minoration (the lower bound in Theorem 5.5 or 5.6) and obtain the following recent important result.

Theorem 5.8. *There is a constant C' such that for all Banach spaces E, for all $n \geq 1$ and for all $u : \ell_2^n \to E$, we have*

$$\sup_{k \geq 1} k^{1/2} c_k(u^*) \leq C'\ell(u).$$

The proof will use several lemmas of independent interest.

Lemma 5.9. *Let $\{g_{ij}\}$ be a double sequence of i.i.d. normal Gaussian random variables on some probability space $(\Omega, \mathcal{A}, \mathbf{P})$. Let k, n be integers. Let $G = (g_{ij})_{\substack{i \leq n \\ j \leq k}}$. We consider the matrix G as a (random) operator from ℓ_2^k into ℓ_2^n. Let $u : \ell_2^n \to E$ be an operator with values in a Banach space E. There is a numerical constant K_1 such that, with probability $> 2/3$, we have*

$$\|uG\|_{B(\ell_2^k, E)} \leq K_1 \left[\ell(u) + k^{1/2}\|u\|\right].$$

Proof: We need to recall some elementary facts from Chapter 4. There is a subset A of the sphere of ℓ_2^k which is a $(1/2)$-net and has cardinality $\operatorname{card}(A) \leq 5^k$ (cf. Lemma 4.10). Then, by Remark 4.12, for any $v : \ell_2^k \to E$, we have

$$(5.28) \qquad \|v\|_{B(\ell_2^k, E)} \leq 2\sup_{x \in A} \|v(x)\|.$$

We will use this fact for $v = uG$. Let

$$X_j = \sum_{i=1}^n g_{ij} u(e_i).$$

Clearly, $X_1, \ldots, X_k$ are independent and have the same distribution as the variable

$$X = \sum_{1}^{n} g_i u(e_i).$$

Consider a point x in A. Recall

$$\sum_{1}^{k} x_j^2 = 1.$$

We have

$$uG(x) = \sum_{1}^{k} x_j X_j,$$

hence $uG(x)$ has the same distribution as X, so that for all $t > 0$

(5.29) $\quad \mathbf{P}\left\{ \|uG(x)\| - \mathbf{E}\|uG(x)\| > t \right\} = \mathbf{P}\left\{ \|X\| - \mathbf{E}\|X\| > t \right\}.$

By Theorem 4.7 we have (since $\sigma(X) = \|u\|$)

$$\mathbf{P}\left\{ \|X\| - \mathbf{E}\|X\| > t \right\} \le 2\exp{-(Kt^2\|u\|^{-2})}.$$

Hence, by (5.29)

(5.30) $\quad \mathbf{P}\left\{ \sup_{x \in A} \|uG(x)\| > \mathbf{E}\|X\| + t \right\} \le 2.5^k \exp(-Kt^2\|u\|^{-2}).$

We may clearly assume $k \ge 2$. There is obviously a numerical constant λ such that

$$2.5^k \exp(-K\lambda^2 k) < \frac{1}{3} \text{ for all } k \ge 2.$$

Then we find that with probability $> \frac{2}{3}$ we have

$$\sup_{x \in A} \|uG(x)\| \le \mathbf{E}\|X\| + \lambda\|u\|k^{1/2};$$

hence by (5.28)

$$\|uG\| \le 2\left(\ell(u) + \lambda\|u\|k^{1/2} \right). \quad \blacksquare$$

Lemma 5.10. *Consider $u : \ell_2^n \to E$ and $k \ge 1$ as above. For any $\varepsilon > 0$, let $B_\varepsilon = B_E + \varepsilon^{-1}u(B_{\ell_2^n})$. We denote by E_ε the space E equipped with the norm for which B_ε is the unit ball. Then, for each $\varepsilon > 0$ we have on a set of probability $> 2/3$*

$$\|uG\|_{B(\ell_2^k, E_\varepsilon)} \le K_1\left[\ell(u) + k^{1/2}\varepsilon \right].$$

Proof: We simply apply Lemma 5.9 and observe that obviously $\|u\|_{B(\ell_2^n, E_\varepsilon)} \le \varepsilon$.

Lemma 5.11. *There are positive numerical constants α and β with the following property: Let k, n and m be arbitrary positive integers. Then, for all $T \subset \ell_2^n$ with* $\text{card}(T) < 2^m$, *there is a subset $\Omega_1 \subset \Omega$ with $\mathbf{P}(\Omega_1) > 3/4$ such that for all ω in Ω_1 we have*

$$\forall\, y \in T \quad \alpha 2^{-m/k} \|y\|_{\ell_2^n} \leq k^{-1/2} \|G^*(\omega)y\|_{\ell_2^k} \leq \left[2 + \beta(m/k)^{1/2}\right] \|y\|_{\ell_2^n}.$$

Proof: We have $G^*y = \sum_{ij} g_{ij} y_i e_j$, so that G^*y has the same distribution as the random variable

$$\left(\sum y_i^2\right)^{1/2} \cdot \left(\sum_1^k g_j e_j\right).$$

Let us denote simply

$$Z^y = k^{-1/2} \left(\|G^*y\|_{\ell_2^k}\right) \left(\|y\|_{\ell_2^n}\right)^{-1}.$$

By the preceding observation, Z^y has the same distribution as the variable

$$Z_k = k^{-1/2} \left(\sum_1^k g_i^2\right)^{1/2}.$$

We claim that there are positive numerical constants α_1 and β_1 such that for all k we have

(5.31) $$\forall\, s > 0 \quad \mathbf{P}\{Z_k < s\} \leq (\alpha_1 s)^k$$

(5.32) $$\forall\, t > 2 \quad \mathbf{P}\{Z_k > t\} \leq \exp -(\beta_1 k t^2).$$

The proof of (5.31) and (5.32) are outlined below. It is easy to derive Lemma 5.11 from these estimates. Indeed, for each fixed non-zero y in ℓ_2^n we have for all $0 < s < t < \infty$ $(t > 2)$

$$\mathbf{P}\{Z_y \notin [s,t]\} = \mathbf{P}\{Z_k \notin [s,t]\}$$
$$\leq (\alpha_1 s)^k + \exp -(\beta_1 k t^2).$$

Hence,

(5.33) $$\mathbf{P}\{\exists y \in T | Z^y \notin [s,t]\} \leq \text{card}(T) \left[(\alpha_1 s)^k + \exp -(\beta_1 k t^2)\right]$$
$$\leq 2^m \left[(\alpha_1 s)^k + \exp -\beta_1 k t^2\right].$$

Therefore, if we choose $s = 8^{-1}2^{-m/k}\alpha_1^{-1}$ and

$$t = 2 + ((m+4)k^{-1}(\mathrm{Log}\ 2)\beta_1^{-1})^{1/2}$$

then we find the probability (5.33) less than $1/4$, hence the complement Ω_1 has probability more than $3/4$ and we obtain Lemma 5.11.

Now, let us justify (5.31) and (5.32). We have

$$\mathbf{P}\{Z_k < s\} = (2\pi)^{-k/2} \int_{\sum x_i^2 < s^2 k} \exp - \left(\frac{1}{2}\right)\left(\sum_1^k x_i^2\right) dx_1 \dots dx_k,$$

hence majorizing the exponential term simply by 1

$$\leq (2\pi)^{-k/2} \left(sk^{1/2}\right)^k \mathrm{vol}\left(B_{\ell_2^k}\right).$$

Since $\mathrm{vol}(B_{\ell_2^k}) \leq (ck^{-1/2})^k$ for some numerical constant c, we find (5.31) with $\alpha_1 = c(2\pi)^{-1/2}$.

We now turn to (5.32). This is a (very) special case of Theorem 4.7 (and the last line of its proof). Let us give a direct argument. We have

$$\mathbf{P}\{Z_k > t\} = (2\pi)^{-k/2} \int_{\sum x_i^2 > t^2 k} \exp - \left(\frac{1}{2}\right)\sum x_i^2\ dx_1 \dots dx_k$$

$$\leq (2\pi)^{-k/2} \exp - \left(4^{-1}t^2 k\right) \int \exp - \left(\frac{1}{4}\sum x_i^2\right) dx_1 \dots dx_k.$$

But the last integral is equal to $(2\pi)^{k/2} \cdot 2^{k/2}$ (make the change of variable $y = 2^{-1/2}x$), so that we obtain

$$\mathbf{P}\{Z_k > t\} \leq 2^{k/2} \exp - \left(4^{-1}t^2 k\right)$$

and (5.32) follows (by choosing for β_1 a small enough numerical value). ∎

We can now complete the proof of Theorem 5.8.

Proof of Theorem 5.8: Consider $u : \ell_2^n \to E$. Let $K = u^*(B_{E^*})$. By the lower bound in Theorem 5.5, we have $e_k(u^*) \leq k^{-1/2}C_1^{-1}\ell(u)$ for all k.

Hence, there is a subset $T \subset \ell_2^n$ such that $\mathrm{card}(T) < 2^k$ and $\forall s \leq \varepsilon_k\ \exists\ t \in T$ satisfying $\|t - s\|_2 \leq \varepsilon_k$ with $\varepsilon_k = k^{-1/2}C_1^{-1}\ell(u)$.

Note that we can assume that $T \subset K$ if we replace ε_k by $2\varepsilon_k$. Thus we assume $T \subset K$. Note that, in the above, $t - s$ will belong to $2K$ so that we have now

(5.34) $\quad \forall s \in K \;\; \exists\, t \in T$ such that $t - s \in (2K) \cap (2\varepsilon_k B_{\ell_2^n})$.

We consider the random $n \times k$ matrix G as above. By the two preceding lemmas, we can find with *positive* probability (actually $> 1/3$) a matrix G satisfying both of the following properties.

(i) $\qquad\qquad\qquad \|uG\|_{B(\ell_2^k, E_{\varepsilon_k})} \leq K_1(\ell(u) + k^{1/2}\varepsilon_k),$

(ii) $\qquad\qquad\qquad \forall\, t \in T \quad \alpha\|t\|_2 \leq k^{-1/2}\|G^*t\|_2.$

Let $F = \mathrm{Ker}\,(G^*u^*)$. Then F is a subspace of E^* with

$$\mathrm{codim}\, F = rk(G^*u^*) \leq k.$$

Moreover, for any ζ in B_F, $u^*(\zeta)$ is in K, hence by (5.34) there is a t in T such that

(5.35) $\qquad\qquad\qquad u^*(\zeta) - t \in 2(K \cap \varepsilon_k B_{\ell_2^n}).$

In particular $\|u^*(\zeta) - t\|_2 \leq 2\varepsilon_k$, but on the other hand since ζ lies in F, we have

$$G^*u^*\zeta = 0,$$

so that (5.35) implies

$$G^*t \in 2G^*(K \cap \varepsilon_k B_{\ell_2^n}) = 2G^*u^*(\widetilde{B})$$

with $\widetilde{B} = B_{E^*} \cap \varepsilon_k u^{*-1}(B_{\ell_2^n})$. But $\widetilde{B} \subset 2(B_{E_{\varepsilon_k}})^\circ$, hence

$$G^*t \in 4G^*u^*(B^\circ_{E_{\varepsilon_k}}).$$

By (i) above, this implies (using $\|G^*u^*\| = \|uG\|$)

$$\|G^*t\|_2 \leq 4K_1(\ell(u) + k^{1/2}\varepsilon_k).$$

By (ii) above, it follows that

$$\|t\|_2 \leq \alpha^{-1}4K_1(\ell(u)k^{-1/2} + \varepsilon_k),$$

and finally by (5.35)

$$\|u^*(\zeta)\|_2 \leq \|u^*\zeta - t\|_2 + \|t\|_2$$
$$\leq 2\varepsilon_k + \alpha^{-1}4K_1(\ell(u)k^{-1/2} + \varepsilon_k)$$
$$\leq K_2\ell(u)k^{-1/2}.$$

Thus we have obtained a numerical constant K_2 such that

$$c_{k+1}(u^*) \leq K_2\ell(u)k^{-1/2}$$

and Theorem 5.8 follows immediately. ∎

Corollary 5.12. *There are positive numerical constants K_3 and K_4 such that for all Banach spaces E and all operators $u : \ell_2 \to E$ we have*

(5.36) $\qquad K_4 \sup k^{1/2} e_k(u) \leq \sup k^{1/2} c_k(u^*) \leq K_3\ell(u).$

Proof: Recall (cf. the Remark after Theorem 5.5) that $\ell(u) < \infty$ implies that u is compact so u is approximable in norm by finite rank operators. Therefore, the inequality $\sup k^{1/2} c_k(u^*) \leq C'\ell(u)$ follows immediately from Theorem 5.8.

On the other hand, we have $d_k(u^{**}) \leq c_k(u^*)$ and by Theorem 5.2

$$\sup k^{1/2} e_k(u^{**}) \leq \rho_{1/2} \sup k^{1/2} d_k(u^{**}),$$

since $e_k(u) \leq e_k(u^{**})$, we obtain the left side of (5.36). ∎

Remark 5.13: To give a concrete application of Theorem 5.8, consider the mapping $i_n : \ell_2^n \to \ell_q^n$ induced by the identity operator of $\mathbf{R}^n$ for $2 \leq q < \infty$. We have seen in Chapter 4 (cf. Lemma 4.14) that

$$\ell(i_n) \leq \gamma(q)n^{1/q},$$

where $\gamma(q) = \|g_1\|_q \leq K\sqrt{q}$ for some numerical constant K. Therefore, by Theorem 5.8 we have

$$d_k(i_n) = c_k(i_n^*) \leq C'\gamma(q)n^{1/q}k^{-1/2};$$

hence since $\|i_n\| \leq 1$

$$d_k(i_n) \leq \min\{1, C'\gamma(q)n^{1/q}k^{-1/2}\}.$$

Gluskin showed in [Gl1] that this bound is optimal, namely that at least if $k \geq n^{2/q}$ then $d_k(i_n)$ is equivalent to $k^{-1/2}\ell(i_n)$. In the case $q = \infty$, i.e. for $i_n : \ell_2^n \to \ell_\infty^n$, the preceding remark yields

$$c_k(i_n^*) = d_k(i_n) \leq c_1 k^{-1/2}(1 + \text{Log } n)^{1/2}$$

for some numerical constant c_1.

Since $\|i_n^{-1}\| = n^{1/2}$ in that case ($q = \infty$), this implies that for every $k \leq n$ there is a subspace $F \subset \ell_1^n$ with $\dim F = d > n - k$ such that

$$d(F, \ell_2^d) \leq c_1(n/k)^{1/2}(1 + \text{Log } n)^{1/2}.$$

It is not clear what is the best possible bound in the last estimate. In the sequel (cf. the Remark before Theorem 10.4) we will study similar questions with a general cotype 2 or weak cotype 2 space in the place of ℓ_1^n.

Actually, in the case of $i_n : \ell_2^n \to \ell_\infty^n$, Garnaev and Gluskin [GG] have proved that $d_k(i_n)$ is equivalent (up to absolute constants) to

$$\min\left\{1, k^{-1/2}\left(\text{Log }\left(1 + \frac{n}{k}\right)\right)^{-1/2}\right\}.$$

The majoration of $d_k(i_n)$ is proved in [GG] by the same method as in [Gl1]. See [PT2] for an alternate proof. Also, for the lower bound, a different proof is given in [CP], based on Theorem 5.2 and the corresponding estimates of $e_k(i_n)$ which were obtained by Schütt [Sc1].

The following result is similar to Theorem 5.8 but is much easier to prove. The reader who wishes to skip the proof of Theorem 5.8 can use this result instead of Theorem 5.8 in the sequel. Its only drawback is that it yields slightly worse quantitative estimates.

Theorem 5.14. *There is a numerical constant C such that for all Banach spaces Y, all n, and all operators $v : Y \to \ell_2^n$ we have for all $k \leq n$ and all integers m*

(5.37) $$c_k(v) \leq C(n/k)^{1/2}2^{m/k}e_m(v),$$

in particular

(5.38) $$c_k(v) \leq 2C(n/k)^{1/2}e_k(v).$$

Remark: Clearly (5.38) implies

(5.39) $$c_k(v) \leq 2C(n/k)S(v)n^{-1/2}$$

where $S(v) = \sup k^{1/2}e_k(v)$.

Therefore, by Theorem 5.5, this implies

(5.40) $$c_k(v) \leq 2CC_1^{-1}(n/k)\ell(v^*)n^{-1/2},$$

which is a weakened form of Theorem 5.8 which can be used as a substitute for Theorem 5.8 throughout Chapters 7 and 8.

Proof of Theorem 5.14: We will use the well-known fact that for all $k \leq n$ we have

$$(5.41) \qquad \mathbf{E}\|G\|_{B(\ell_2^k, \ell_2^n)} \leq \alpha_2 n^{1/2}$$

for some numerical constant α_2 (independent of k or n). Indeed, this can easily be derived from (4.16) and the fact that we can estimate the norm in $B(\ell_2^k, \ell_2^n)$ using a $(1/2)$-net in the unit balls of ℓ_2^k and ℓ_2^n (cf. Remark 4.12). We leave the details as an exercise for the reader.

We then use only Lemma 5.11 to complete the proof. Let $\varepsilon = e_k(v)$. There is a subset T in ℓ_2^n with $\mathrm{card}(T) \leq 2^k - 1$ which is an ε-net for $v(B_Y)$. Let us denote simply by $|\ |_2$ the norm in the space ℓ_2^k.

By (5.41), we have with probability $> 2/3$

$$(5.42) \qquad \|G\|_{B(\ell_2^k, \ell_2^n)} \leq 3\alpha_2 n^{1/2}.$$

Hence, by Lemma 5.11, we have with *positive* probability both (5.42) and

$$(5.43) \qquad \forall\, t \in T \quad \alpha 2^{-m/k}|t|_2 \leq k^{-1/2}|G^* t|_2.$$

Now fix a matrix G^* satisfying (5.42) and (5.43) and let $F = \mathrm{Ker}(G^* v)$.

Clearly, $\mathrm{codim}\, F \leq k$. We have $\forall\, \zeta \in B_F\ \exists\, t \in T$ such that

$$(5.44) \qquad |v(\zeta) - t|_2 \leq \varepsilon.$$

Hence by (5.42) $|G^* v(\zeta) - G^* t|_2 \leq 3\alpha_2 n^{1/2}\varepsilon$. But since $\zeta \in F$ implies $G^* v(\zeta) = 0$, this yields $|G^* t|_2 \leq 3\alpha_2 n^{1/2}\varepsilon$, hence by (5.43)

$$|t|_2 \leq 3\alpha_2 \alpha^{-1} 2^{m/k}(n/k)^{1/2}\varepsilon.$$

Thus we conclude by (5.44):

$$|v(\zeta)|_2 \leq \varepsilon + 3\alpha_2 \alpha^{-1} 2^{m/k}(n/k)^{1/2}\varepsilon,$$

or

$$\|v_{|F}\| \leq \varepsilon + 3\alpha_2 \alpha^{-1} 2^{m/k}(n/k)^{1/2}\varepsilon.$$

This yields an upper bound for $c_{k+1}(v)$ and the announced result follows. ∎

Remark 5.15: Let $T : \ell_2 \to \ell_2$ be a compact operator. Let us denote by $\{\lambda_n\}$ the eigenvalues of $|T|$ arranged as usual in non-increasing order and repeated according to their multiplicity. Then we can compute (up to equivalence) the entropy numbers of T as follows: Let

$$\phi_n(T) = \sup_{k \geq 1} 2^{-n/k} \, |\Pi_{1 \leq j \leq k} \lambda_j|^{1/k} \, .$$

We have (for all $n \geq 1$)

(5.45) $$\phi_n(T) \leq e_{n+1}(T) \leq 6\phi_n(T).$$

Let us briefly sketch the proof of (5.45). We may as well assume that T is a diagonal operator relative to an orthonormal basis denoted by (e_n). Let H_k be the span of $(e_1, \dots, e_k)$ and let $T_k : H_k \to H_k$ be the restriction of T to H_k. Let B_k denote the unit ball of H_k.

The left side of (5.45) is immediate; we have

$$|\Pi_1^k \lambda_j| = |\det(T_k)| = \frac{\text{vol}(TB_k)}{\text{vol}(B_k)} \leq 2^n e_{n+1}(T_k)^k \leq 2^n e_{n+1}(T)^k.$$

Hence, taking kth roots, the left side of (5.45) follows.

To prove the other side, let us assume for simplicity that $\phi_n(T) \leq 1$. Let m be the first integer such that $\lambda_{m+1} \leq 2$. If $m = 0$ (i.e., $\lambda_1 \leq 2$), then $e_n(T) \leq \|T\| \leq 2$ and we are done, so we may assume $m \geq 1$.

Note that $\lambda_1 \geq \dots \geq \lambda_m \geq 2$ implies (by unconditionality)

$$2B_m \subset T_m(B_m)$$

hence for all $t > 0$

$$tT_m(B_m) + B_m \subset \left(t + \frac{1}{2}\right) T_m(B_m).$$

Using (4.21) (from Lemma 4.16), we find (letting $t = \frac{1}{2}$)

$$N(T_m(B_m), 4B_m) \leq \frac{\text{vol}((\frac{1}{2})T_m(B_m) + B_m)}{\text{vol}(B_m)}$$

$$\leq \frac{\text{vol}(T_m(B_m))}{\text{vol}(B_m)} = |\Pi_1^m \lambda_j|$$

and this is no more than 2^n since $\phi_n(T) \leq 1$. It follows that $e_{n+1}(T_m) \leq 4$.

Since $\|T - T_m\| \leq \lambda_{m+1} \leq 2$, we find

$$e_{n+1}(T) \leq e_{n+1}(T_m) + \|T - T_m\|$$
$$\leq 6.$$

By homogeneity this proves the right side of (5.45).

The preceding argument is taken from [GKS] and works equally well if T is a diagonal operator with coefficients (λ_n) on a Banach space with a 1-unconditional basis.

Finally, we observe that, by the polar decomposition of T (say $T = \cup|T|$ and $|T| = \cup^* T$ with $\| \cup \| \leq 1$) we have

$$e_{n+1}(T) \leq e_{n+1}(|T|) \leq e_{n+1}(T);$$

therefore, in particular,

(5.46) $$e_{n+1}(T^*) = e_{n+1}(T).$$

Notes and Remarks

The numbers $(a_n(u)), (c_n(u))$ and $(d_n(u))$ have been extensively studied in the literature, especially in Approximation Theory (the numbers $d_n(u)$ are called there Kolmogorov's n-widths). Of course the covering numbers $N(K_1, \varepsilon K_2)$ have also been considered for a long time (cf. e.g. [Ko], [KT]), but the entropy numbers $(e_n(u))$ were formally introduced by Pietsch (cf. [Pi1]) primarily motivated by some earlier work of Boris Mityagin [Mi] (cf. also [MiP]). Pietsch made the first attempts to compare $(e_n(u))$ with the other s-numbers.

Proposition 5.1 is elementary.

Theorem 5.2 is due to B. Carl [Ca1], as well as Theorem 5.3. See also [CT].

Corollary 5.4 essentially goes back to the early paper of Mityagin [Mi] who studied the entropy numbers of nuclear and Hilbert–Schmidt operators on ℓ_2.

Theorems 5.5 and 5.6 are due to Dudley [Du1] and Sudakov [Su] More precisely, Dudley [Du1] proved the upper bound in (5.13) and conjectured the lower bound which was later proved by Sudakov [Su]. Theorem 5.5 is merely a reformulation of Theorem 5.6 in the language of linear operators. The reader should note that Sudakov minoration (i.e. the lower bound in Theorem 5.5) is closely related to the Urysohn inequality. Indeed, for all $u : \ell_2^n \to X$ we have obviously

$$\left(\frac{\text{vol}(u^*B_{X^*})}{\text{vol}(B_{\ell_2^n})}\right)^{1/n} \leq 2e_n(u^*) \leq 2C_1^{-1}n^{-1/2}\ell(u)$$

(the latter by Theorem 5.5). This is analogous (up to the numerical constant) to the Urysohn inequality (Corollary 1.4) if one takes (0.5) into account.

In connection with Theorem 5.5 and 5.6, it is natural to ask for an *equivalent* of the ℓ-norm of an operator $u : \ell_2^n \to E$ from ℓ_2^n into a Banach space E. This remained for a long time an open problem although it was well known that none of the inequalities appearing in Theorem 5.5 can be reversed in general. This major problem was solved recently by Talagrand [Ta] who obtained a two-sided bound for the expectation of the supremum of a Gaussian process (as in Theorem 5.6) in terms of "majorizing measures" instead of entropy numbers. More precisely, Talagrand showed that $\ell(u)$ is equivalent to another norm $\tilde{\ell}(u)$ which can be described as follows. Let $\Delta : c_o \to c_o$ be the diagonal operator with diagonal coefficients

$$\left\{(1 + \text{Log } n)^{1/2} \,|n = 1, 2, \dots\right\}.$$

Let S_1, S_2 be subspaces of c_o such that $\Delta(S_1) \subset S_2$, and let $\tilde{\Delta} : S_1 \to S_2$ denote the restriction of Δ. Let $u : \ell_2^n \to E$ be as above. Let us assume that there are operators $B : \ell_2^n \to S_1$ and $A : S_2 \to E$ such that $u = A\tilde{\Delta}B$. Then we may define

$$\tilde{\ell}(u) = \inf\{\|A\| \, \|B\|\},$$

where the infimum runs over all possible subspaces S_1, S_2 and all possible factorizations of the above form. By a deep theorem of Talagrand [Ta] there is an absolute constant C such that for all n, for all Banach spaces E, and for all $u : \ell_2^n \to E$ we have

$$\frac{1}{C}\tilde{\ell}(u) \leq \ell(u) \leq C\tilde{\ell}(u).$$

The important part is the first inequality (the second one is easy by standard arguments).

In Lemma 5.7, (5.16) is due to Slepian [Sl] but the first part (without the factor 2) is due to Sudakov who states it without proof in [Su].

For a proof of Lemma 5.7 (without the extra 2) we refer the reader
to [BC] or [Fe1]. There is also a *geometric* proof of this lemma due to
Gromov; cf. [Gr]. More recently Kahane gave a quicker proof of this;
see [Kah1]. See [To] for related inequalities.

Kahane's argument also yields a very direct proof of the following generalization of Slepian's Lemma due to Y. Gordon [Go1]. Let
$(X_{ij})(Y_{ij})$ be two centered Gaussian processes (indexed by couples i, j
with $1 \leq i \leq n, 1 \leq j \leq m$) satisfying the following:

$$\mathbf{E}X_{ij}^2 = \mathbf{E}Y_{ij}^2$$
$$\|X_{ij} - X_{ik}\|_2 \geq \|Y_{ij} - Y_{ik}\|_2 \quad \forall \, i, j, k$$
$$\|Y_{ij} - Y_{\ell k}\|_2 \leq \|X_{ij} - X_{\ell k}\|_2 \quad \forall \, i \neq \ell \, \forall \, j, k.$$

Then for all real numbers c_{ij}

$$\mathbf{P}\left(\cap_i \cup_j \{X_{ij} \geq c_{ij}\}\right) \geq \mathbf{P}\left(\cap_i \cup_j \{Y_{ij} \geq c_{ij}\}\right).$$

In particular, for all $c > 0$,

$$\mathbf{P}\left(\inf_i \sup_j X_{ij} \geq c\right) \geq \mathbf{P}\left(\inf_i \sup_j Y_{ij} \geq c\right).$$

Using this lemma, Gordon has given a very direct proof of Dvoretzky's
Theorem. More recently, in [Go2] he observed that this lemma also
gives a direct proof of Theorem 5.8 (which first appeared in [PT1]).
Although this approach is clearly more direct and elegant (especially
now that Kahane's proof is available), we feel that the more traditional approach which we followed to prove Dvoretzky's Theorem and
Theorem 5.8 is more *instructive* for the reader.

Theorem 5.8 is due to Pajor and Tomczak-Jaegermann [PT1]. It
improved an earlier result of Milman [M4]. For the qualitative results
in subsequent chapters, we could use either Milman's original estimate
(or actually Theorem 5.14) instead of Theorem 5.8, but we choose
to use Theorem 5.8 since it gives better quantitative estimates, for
instance in the QS-Theorem and the estimate of $d_X(\delta)$ in Chapter 10.

Theorem 5.8 may be viewed as a refinement of the lower bound in
Theorem 5.5 if we take into account Theorem 5.2.

The proof of Theorem 5.8 is based on ideas from a remarkable
paper of Gluskin [Gl1]. Actually, the proof of Theorem 5.8 in [PT1]
(or above) follows closely the proof of Theorem 3 in [Gl1] (see also
Remark 2 in [Gl1]) but uses the Sudakov minoration instead of the

more concrete estimates (for ℓ_p^n balls) used by Gluskin in his paper. See also [Gl2] for related information.

Let us denote by U an orthogonal matrix in $O(n)$ equipped with its normalized Haar measure. Lemmas 5.9, 5.10, and 5.11 appear in [PT1] with $\sqrt{n}U$ in the place of G. It is well known that both cases are similar and often can be proved with similar methods cf. e.g. [MPi]. We chose to use Gaussian variables. As before in Chapter 4, we feel that this is simpler, but it is really very much a matter of taste. We should mention in passing that Carl and Pajor have proved an analogue of Theorem 5.8 for random choices of signs instead of Gaussian variables. Namely, they prove in [CP] that for all Banach spaces E and for all $u : \ell_2^n \to E$ we have

$$\forall\, k \le n \quad c_k(u^*) \le C k^{-1/2} \left(1 + \operatorname{Log} \frac{n}{k}\right)^{1/2} \mathbf{E} \left\| \sum_1^n \varepsilon_i u(e_i) \right\|,$$

where C is a numerical constant and where $\varepsilon_1, \dots, \varepsilon_n$ is a sequence of independent random variables taking the values $+1$ and -1 with equal probability $1/2$.

Lemma 5.11 has appeared already in various papers in one form or another but I am grateful to A. Pajor for communicating it to me in the form given above.

Corollary 5.12 is due to Pajor–Tomczak [PT1] but the left side of (5.36) is a particular case of Theorem 5.2, hence it is due to Carl [Ca1]. Finally, (5.39) and a weaker form of Theorem 5.14 appeared (with essentially the same proof) in the appendix of [MP1]. Again, I am indebted to A. Pajor for showing me the stronger formulation of Theorem 5.14 given above.

I believe that Remark 5.15 and (5.45) essentially go back to Mityagin in the Hilbert space case. The proof included in Remark 5.15 is taken from [GKS], where the case of diagonal operators on a space with unconditional basis is treated.

Concerning entropy numbers, the so-called *duality problem* for entropy numbers is the following set of questions (probably due to Carl).

(i) Is there a constant c such that for all compact operators $u : X \to Y$ betweeen Banach spaces we have

$$\forall\, n \ge 1 \quad e_{[cn]}(u^*) \le c\, e_n(u)?$$

(ii) Is it true that $(e_n(u))_n$ is in ℓ_p iff $e_n(u^*)$ is in ℓ_p?

Of course a positive answer to (i) implies a positive answer to (ii).
By (5.46), the answer to (i) and (ii) is positive if X and Y are *both*
Hilbert spaces.

These problems have been intensively investigated in recent months
but they are still open in full generality.

In Chapter 8 (Corollary 8.11) we include a result of König–Milman
limited to operators of rank n. There are partial results on these
questions in the papers [PT2] and [PT3]. Recently, N. Tomczak-
Jaegermann has obtained a positive answer to (ii) in the case when
X or Y is a Hilbert space; cf. [TJ2]. For more information see [Ca3],
[GKS], [Sc1].

Chapter 6

Volume Ratio

The notion of volume ratio originates in the work of Kašin [Ka1] which has important applications in approximation theory. It was formally introduced by Szarek in the paper [S] (cf. also [ST]) which we follow in this chapter. Let E be an n-dimensional normed space with unit ball $B_E \subset \mathbf{R}^n$. We define the *volume ratio* of E as the number

$$vr(E) = \inf \left\{ \left(\frac{\text{vol}(B_E)}{\text{vol}(D)} \right)^{1/n} \right\},$$

where the infimum runs over all ellipsoids $D \subset B$. Equivalently, if $D_E^{\max} \subset B_E$ is the ellipsoid of maximal volume we have

$$vr(E) = \left(\frac{\text{vol}(B_E)}{\text{vol}(D_E^{\max})} \right)^{1/n}.$$

For example, we have trivally $vr(\ell_2^n) = 1$ and this characterizes obviously ℓ_2^n isometrically. But it is less trivial (and much more significant) that there is a numerical constant C_1 such that $vr(\ell_1^n) \leq C_1$ for all $n \geq 1$. Indeed, let $D = n^{-1/2} B_{\ell_2^n}$. Then $D \subset B_{\ell_1^n}$ and by (1.17)

$$\text{vol}(B_{\ell_2^n}) = \pi^{n/2} \Gamma \left(\frac{n}{2} + 1 \right)^{-1},$$

while

$$\text{vol}(B_{\ell_1^n}) = 2^n (n!)^{-1}.$$

Therefore,

$$(6.1) \qquad vr(\ell_1^n) = \left(\frac{\text{vol}(B_{\ell_1^n})}{\text{vol}(D)} \right)^{1/n} \leq \left(\frac{2e}{\pi} \right)^{1/2}.$$

(Note that by a symmetry argument D is indeed the maximal volume ellipsoid for $B_{\ell_1^n}$.) Moreover, a simple computation shows similarly for $1 \leq p \leq 2$,

$$vr(\ell_p^n) = \left(\frac{\text{vol}(B_{\ell_p^n})}{\text{vol} \left(n^{1/2 - 1/p} B_{\ell_2^n} \right)} \right)^{1/n} \leq vr(\ell_1^n).$$

Finally, we observe that $vr(E)$ depends only on the space E up to isometry and not on the particular realization of E in $\mathbf{R}^n$. More precisely, we have, if F is another n-dimensional normed space,

$$vr(E) \leq d(E, F)vr(F).$$

The main result of this chapter is the following.

Theorem 6.1.   *Let B be a ball in $\mathbf{R}^n$ and let D be an ellipsoid such that $D \subset B$.*
   Let

$$A = \left(\frac{\mathrm{vol}(B)}{\mathrm{vol}(D)} \right)^{1/n} .$$

We denote by $\| \ \|$ the gauge of B and by $| \ |$ the gauge of D so that $\|x\| \leq |x|$ for all x in $\mathbf{R}^n$.
   Then for each $k = 1, 2, \ldots, n-1$ there is a subspace $F \subset \mathbf{R}^n$ with $\dim F = k$ such that

$$\forall \, x \in F \quad |x| \leq (4\pi A)^{\frac{n}{n-k}} \|x\|.$$

Moreover, if we denote by m the canonical probability measure on the Grassman manifold G_{nk} of all k-dimensional subspaces of $\mathbf{R}^n$ (equipped with the scalar product associated to D) then the set Ω_0 of all the subspaces F as above satisfies

$$m(\Omega_0) > 1 - 2^{-n}.$$

Remark: By a change of algebraic basis of $\mathbf{R}^n$, we may as well assume that $D = B_{\ell_2^n}$ in Theorem 6.1. Let us denote by ν the canonical normalized Haar measure on the orthogonal group $O(n)$, and let F_o be a fixed k-dimensional subspace, for instance $F_0 = [e_1, \ldots, e_k]$. Clearly, any F in G_{nk} can be written as $T(F_o)$ for some T in $O(n)$. Then for any measurable subset Ω of the set G_{nk} of all k-dimensional subspaces of $\mathbf{R}^n$, we have

$$m(\Omega) = \nu(\{T \in O(n), T(F_o) \in \Gamma\}).$$

Indeed, m can be characterized as the unique probability measure on G_{nk} which is invariant under the action of $O(n)$.

Proof of Theorem 6.1: Without loss of generality, we may and do assume that $D = B_{\ell_2^n}$. We will need several steps.

Step 1. We have

$$\text{(6.2)} \qquad \frac{\text{vol}(B)}{\text{vol}(D)} = \int_S \|x\|^{-n} \sigma(dx),$$

where S is the Euclidean unit sphere of $\mathbf{R}^n$ and σ its normalized area measure.

Indeed, passing in polar coordinates, we find

$$\text{vol}(B) = \lambda_n \int_S \sigma(dx) \int_0^{\frac{1}{\|x\|}} n r^{n-1}\, dr,$$

where λ_n is a fixed normalizing constant independent of B. In particular, taking $B = D$, we find necessarily $\lambda_n = \text{vol}(D)$. This yields (6.2).

Step 2. Let m be as above. For any $F \subset \mathbf{R}^n$ with $\dim F = k$, let σ_F be the normalized measure on the Euclidean unit sphere S_F of F. Then for any measurable function $\varphi : S \to \mathbf{R}$ we have

$$\int_S \varphi(x)\, \sigma(dx) = \int_{G_{nk}} m(dF) \int_{S_F} \varphi(x)\, \sigma_F(dx).$$

This is immediate since σ is the unique probability on S invariant under rotations.

Step 3. Let $x \in S_F, F \subset \mathbf{R}^n, \dim F = k$. Then for all $0 < \delta < \sqrt{2}$ we have

$$\sigma_F(\{y \in S_F, |x - y| \le \delta\}) > \left(\frac{\delta}{\pi}\right)^k.$$

Indeed, this measure is equal to $\frac{F(\alpha)}{F(\pi)}$ with $F(s) = \int_0^s (\sin t)^{k-2}\, dt$ and with α determined by the equality $\frac{\delta}{2} = \sin(\frac{\alpha}{2}), 0 \le \alpha \le \frac{\pi}{2}$.

The reader can check this easily by referring to the picture.

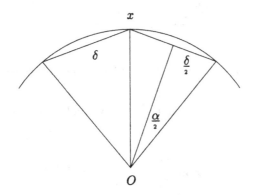

Then we note that $F(\pi) \leq \pi$ and (assuming $k > 1$)

$$F(\alpha) \geq \int_0^\alpha (\sin t)^{k-2} \cos t \, dt = (k-1)^{-1} (\sin \alpha)^{k-1};$$

hence

$$F(\alpha) \geq (k-1)^{-1} \left(2 \sin \frac{\alpha}{2} \cos \frac{\alpha}{2} \right)^{k-1}$$

$$= (k-1)^{-1} \left(\delta \left(1 - \frac{\delta^2}{4} \right)^{1/2} \right)^{k-1}$$

$$\geq (k-1)^{-1} \left(\frac{\delta}{\sqrt{2}} \right)^{k-1},$$

so that we have

$$\frac{F(\alpha)}{F(\pi)} \geq \pi^{-1} (k-1)^{-1} \left(\frac{\delta}{\sqrt{2}} \right)^{k-1} \geq \left(\frac{\delta}{\pi} \right)^k \left(\frac{\pi}{\sqrt{2}} \right)^{k-1} \frac{1}{k-1}$$

$$\geq \left(\frac{\delta}{\pi} \right)^k \frac{2^{k-1}}{k-1} \geq \left(\frac{\delta}{\pi} \right)^k.$$

This establishes Step 3.

Let us now complete the proof.

Let $\Omega_0 = \left\{ F \mid \int_F \|x\|^{-n} \sigma_F(dx) \leq (2A)^n \right\}$.

By Steps 1 and 2, and Markov's inequality, we have

$$m(\Omega_0) > 1 - 2^{-n}.$$

Consider F in Ω_0. Again, by Markov's inequality,

$$(6.3) \qquad \forall r > 0 \quad \sigma_F(\{y \in S_F, \|y\| \le r\}) \le (2Ar)^n.$$

Now we choose r so that

$$(6.4) \qquad (2Ar)^n = \left(\frac{\frac{r}{2}}{\pi}\right)^k.$$

Let $\delta = \frac{r}{2}$. Then the set $L_r = \{y \in S_F | \; \|y\| \le r\}$ cannot contain (by (6.3) and Step 3) any *ball* of radius δ induced on S_F by the Euclidean metric.

In other words, $\forall\, x \in S_F \; \exists\, y \in S_F - L_r$ such that $|x - y| \le \delta$. A *fortiori*, we have $\|x - y\| \le \delta = \frac{r}{2}$, hence

$$\|x\| \ge \|y\| - \|x - y\|$$
$$\ge r - \frac{r}{2} = \frac{r}{2}.$$

Thus we have shown (by homogeneity)

$$\forall\, x \in S_F \quad \|x\| \ge \frac{r}{2}|x|.$$

Finally, we analyze the choice of r made in (6.4). We have $(2Ar)^n = \left(\frac{r}{2\pi}\right)^k$; hence $r^{n-k} = \left(\frac{1}{2A}\right)^n \left(\frac{1}{2\pi}\right)^k$, so that $\left(\frac{r}{2}\right)^{n-k} \ge \left(\frac{1}{2A}\right)^n \left(\frac{1}{2\pi}\right)^n$. Therefore, we conclude $\frac{r}{2} \ge (4\pi)^{-\frac{n}{n-k}}$. ∎

Corollary 6.2. *Let E be an n-dimensional normed space. Then for any $k = 1, 2, \ldots, n-1$, E contains a subspace F with $\dim F = k$ such that*

$$d(F, \ell_2^k) \le (4\pi vr(E))^{\frac{n}{n-k}}.$$

This is an immediate consequence of Theorem 6.1.

Remark: We can also reformulate Theorem 6.1 for operators $u : E \to \ell_2^n$. Note that

$$\left(\frac{\text{vol}(uB_E)}{\text{vol}(B_{l_2^n})}\right)^{1/n} \le 2e_n(u).$$

Hence we obtain from Theorem 6.1 the following inequality:

$$c_{k+1}(u) \le \left(4\pi \frac{\text{vol}(uB_E)}{\text{vol}(B_{l_2^n})}\right)^{\frac{n}{n-k}}$$
$$\le (8\pi e_n(u))^{\frac{n}{n-k}}.$$

In the particular case when u is a diagonal operator from ℓ_2^n into itself, it is easy to see that this cannot be significantly improved.

Corollary 6.3. *Assume $n = 2k$ in the situation of Theorem 6.1. Then there is a decomposition $\mathbf{R}^n = E_1 \oplus E_2$ with E_1, E_2 orthogonal with respect to the inner product structure associated to D, $\dim E_1 = \dim E_2 = k$ and E_1, E_2 satisfy*

$$\forall x \in \ E_1 \cup E_2 \quad C'|x| \leq \|x\| \leq |x|$$

where $C' = (4\pi A)^{-2}$.

Proof: We may as well assume $D = B_{\ell_2^n}$. Note that the map $F \to F^\perp$ preserves the measure m on G_{nk} when $n = 2k$. Therefore

$$m\{F, F \in \Omega_0\} = m\{F, F^\perp \in \Omega_0\} > \frac{1}{2},$$

which imply $m(\{F, F \in \Omega_0 \text{ and } F^\perp \in \Omega_0\}) > 0$. Letting $E_1 = F$ and $E_2 = F^\perp$ with F and $F^\perp$ in Ω_0, we obtain the announced result. ∎

In particular, recalling (6.1) we obtain the following:

Corollary 6.4. *Let us denote by L_p^n the space $\mathbf{R}^n$ equipped with the norm*

$$\|x\|_p = \left(\frac{1}{n} \sum_1^n |x_i|^p \right)^{1/p}.$$

Then if $n = 2k$ there is an orthogonal decomposition $L_2^n = E_1 \oplus E_2$ with $\dim E_1 = \dim E_2 = k$ such that

$$\forall x \in \ E_1 \cup E_2 \quad \|x\|_1 \leq \|x\|_2 \leq C''\|x\|_1,$$

with $C'' = 32e\pi$.

Proof: Indeed, by (6.1) we have

$$\left(\frac{\mathrm{vol}(B_{L_1^n})}{\mathrm{vol}(B_{L_2^n})} \right)^{1/n} \leq \left(\frac{2e}{\pi} \right)^{1/2},$$

so that this is a particular case of the preceding corollary.

This rather striking decomposition of L_1^n can be given an even more *concrete* form as follows.

Corollary 6.5. *There is an absolute constant $C_3 > 0$ such that for each k there is a $2k \times 2k$ orthogonal matrix A satisfying $A^2 = I$ and such that*

$$\forall x \in L_2^{2k} \quad C_3 \|x\|_2 \leq \frac{1}{2}(\|x\|_1 + \|Ax\|_1) \leq \|x\|_2.$$

Proof: The right side is trivial. Let P_1, P_2 be the orthogonal projections onto E_1, E_2.

Let $A = P_1 - P_2$; then we have

$$\frac{1}{2}(\|x\|_1 + \|Ax\|_1) \geq \max(\|P_1 x\|_1, \|P_2 x\|_1)$$
$$\geq (C'')^{-1} \max(\|P_1 x\|_2, \|P_2 x\|_2)$$
$$\geq 2^{-\frac{1}{2}}(C'')^{-1}\|x\|_2. \quad \blacksquare$$

Remark: There is no *constructive* proof of a matrix A satisfying the above. Note that the proof shows actually that we can obtain (for some constant C_3) a set of matrices A with *large* measure in $O(2k)$.

This result has a rather striking infinite-dimensional application obtained by Krivine [Kr] and independently by Kašin [Ka2]. We follow Kašin's argument.

Corollary 6.6. *Let $L_p = L_p([0,1])$. There is an orthogonal decomposition $L_2 = E^1 \oplus E^2$ such that the L_2 and L_1 norms are equivalent both on E_1 and on E_2.*

Proof: We use the Haar orthonormal basis of L_2, which we denote by $\{h_n, n \geq 1\}$. Here $h_1 \equiv 1$, and for all $m \geq 0, 1 \leq j \leq 2^m$ the function h_{2^m+j} is supported by $[(j-1)2^{-m}, j2^{-m}]$ and takes the value $+2^{\frac{m}{2}}$ on the left half of this interval and $-2^{\frac{m}{2}}$ on the right half. Let E_0 be the span of h_1 and let E_{m+1} be the span of $\{h_n, 2^m < n \leq 2^{m+1}\}(m \geq 0)$. Then E_m can be identified with $L_2^{2^{m-1}}$ and E_m equipped with the L_1 norm can be identified naturally with $L_1^{2^{m-1}}$.

This implies that there is for all m an orthogonal decomposition $E_m = E_m^1 \oplus E_m^2$ such that

$$(6.5) \qquad \forall x \in E_m^1 \cup E_m^2 \quad \|x\|_1 \leq \|x\|_2 \leq C''\|x\|_1.$$

Let $E^1 = \underset{m}{\oplus} E_m^1$ and $E^2 = \underset{m}{\oplus} E_m^2$. Let P_m be the orthogonal projection from L_2 onto E_m.

By a classical result in martingale theory, if $1 < p \leq 2$ there is a constant C_p such that

$$(6.6) \qquad \forall\, x \in L_p \quad \left(\sum_{m \geq 0} \|P_m x\|_p^2 \right)^{1/2} \leq C_p \|x\|_p.$$

(Indeed this follows from the unconditionality of the Haar system, c.f. e.g. [Bu].)

Now (6.5) implies

$$\forall\, x \in E^1 \cup E^2 \ \|x\|_2 \leq C'' \|x\|_p \ \text{ if } 1 \leq p \leq 2,$$

hence (6.6) yields for all x either in E^1 or in E^2

$$\|x\|_2 \leq C'' C_p \|x\|_p,$$

and if $1 < p < 2, \frac{1}{p} = \frac{1-\theta}{1} + \frac{\theta}{2}, 0 < \theta < 1$ we have

$$(C'' C_p)^{-1} \|x\|_2 \leq \|x\|_p \leq \|x\|_1^{1-\theta} \|x\|_2^{\theta} ;$$

hence $\|x\|_2 \leq (C'' C_p)^{\frac{1}{1-\theta}}.$ ∎

Note that E^1 and E^2 are necessarily both *infinite dimensional* in the preceding statement.

Remark: We can reformulate the preceeding statement as the existence of an orthonormal basis (φ_n) of L_2 such that the L_2 and L_1 norms are equivalent on the spans of $\{\varphi_n, n \text{ even}\}$ and $\{\varphi_n, n \text{ odd}\}$. No explicit construction of such a system is known. (It is known, however, that the trionometric system fails this.)

Remark: The reader may have noticed that a large part of Theorem 6.1 can be deduced from Theorem 5.14. Indeed, let A be as in Theorem 6.1 and let v be the identity operator on $\mathbf{R}^n$ considered as acting from $\mathbf{R}^n$ equipped with the unit ball B into $\mathbf{R}^n$ equipped with the unit ball D. By (4.22) we have $N(B, D) \leq (3A)^n$, hence we can apply (5.37) with $m = n([\text{Log } 3A] + 1)$ and we obtain for some numerical constant C'

$$\forall\, k \leq n \quad c_k(v) \leq (C' A)^{n/k}.$$

This is essentially (up to the value of the constant C') the first part of Theorem 6.1.

Notes and Remarks

The results on volume ratio in Chapter 6 have been known through the work of Szarek who introduced the notion of volume ratio, cf. [S], [ST]. However, Szarek's work was based on fundamental work of Kašin which had important applications in Approximation Theory.

Thus, Theorem 6.1, Corollaries 6.2 and 6.3, are due to Szarek [S] but were influenced by Kašin's work [Ka1]. Our exposition follows closely Szarek's. Corollary 6.4 (which is sometimes called the Kašin decomposition or splitting of ℓ_1^n) is due to Kašin [Ka1]. As mentioned in the text, Corollary 6.5 was proved independently in [Kr] and [Ka2]. For more information on volume ratio, we refer the reader to [ST], [Pe], [Sc2], [GR], [StR2] and [Ro].

Chapter 7

Milman's Ellipsoids and the
Inverse Brunn–Minkowski Inequality

In this chapter we give a simpler proof (actually two proofs) of several results of V. Milman. The main advantage of our approach is that we prove directly the main result below (Theorem 7.1) without using the previous results of [M1] or [BM] so that we can derive these results from Theorem 7.1. We thus obtain as immediate consequences the *quotient of a subspace* (QS for short) Theorem [M1], the *inverse Santaló inequality* [BM], the *duality of entropy numbers* [KM] (see the next chapter for this) and the *inverse Brunn–Minkowski inequality*. In the next chapter we give another approach: we prove the QS Theorem first and then deduce the inverse Santaló inequality and the duality of entropy numbers as simple consequences, but this alternate approach apparently does not yield Theorem 7.1.

In the sequel, any closed, convex, symmetric subset of $\mathbf{R}^n$ with non-empty interior will be called simply a ball.

Let B, D be subsets of $\mathbf{R}^n$. We denote by B° the polar of B

$$B^\circ = \{x \in \mathbf{R}^n | < x, y > \leq 1 \quad \forall \, y \in B\}.$$

If B is the unit ball of a normed space E, then clearly B° can be identified with the unit ball of the dual E^*.

We define the number

$$(7.1) \qquad M(B, D) = \left(\frac{\mathrm{vol}(B + D)}{\mathrm{vol}(B \cap D)} \cdot \frac{\mathrm{vol}(B^\circ + D^\circ)}{\mathrm{vol}(B^\circ \cap D^\circ)} \right)^{1/n}.$$

We now state the main result of this chapter, which is due to Milman [M5].

Theorem 7.1. *There is a numerical constant C such that, for all $n \geq 1$, for any ball $B \subset \mathbf{R}^n$ there is an ellipsoid D such that*

$$(7.2) \qquad M(B, D) \leq C.$$

99

Any ellipsoid D satisfying (7.2) will be called a Milman ellipsoid associated to B.

Let $u : \mathbf{R}^n \to \mathbf{R}^n$ be a linear invertible transformation. Then clearly

$$M(B, D) = M(u(B), u(D)),$$

so that $M(B, D)$ is an affine invariant of the couple (B, D). A related affine invariant was studied by Santaló ([Sa1]); (cf. also Blaschke's paper [Bl]) we will denote it by

$$s(B) = (\mathrm{vol}(B)\,\mathrm{vol}(B^\circ))^{1/n}.$$

Here again $s(u(B)) = s(B)$, and in particular for any ellipsoid $D \subset \mathbf{R}^n$ we have

(7.3) $$s(D) = s(B_{\ell_2^n}) = (\mathrm{vol}(B_{\ell_2^n}))^{2/n}.$$

Santaló ([Sa1]) proved that $s(B) \le s(B_{\ell_2^n})$ with equality only if B is an ellipsoid.

Recently, using methods from the local theory of Banach spaces such as those described in Chapters 2 and 3 above, Bourgain and Milman have obtained a converse to Santaló's inequality $s(B) \ge cs(B_{\ell_2^n})$ for some absolute constant $c > 0$. Clearly, we can deduce from Theorem 7.1 Santaló's inequality (but without the sharp constant 1) and the inverse inequality of Bourgain–Milman, as follows

Corollary 7.2.   *For all $n \ge 1$, any ball $B \subset \mathbf{R}^n$ satisfies*

$$C^{-1} \le \left(\frac{\mathrm{vol}(B)\mathrm{vol}(B^\circ)}{\mathrm{vol}(B_{\ell_2^n})^2} \right)^{1/n} \le C.$$

Proof: By Theorem 7.1 we have clearly

$$C^{-1}s(D) \le s(B) \le Cs(D)$$

and by (7.3) this implies Corollary 7.2. ∎

The classical Brunn–Minkowski inequality states that if A_1, A_2 are two compact (or suitably measurable) subsets of $\mathbf{R}^n$ we have (see Theorem 1.1)

$$(\mathrm{vol}(A_1))^{1/n} + (\mathrm{vol}(A_2))^{\frac{1}{n}} \le (\mathrm{vol}(A_1 + A_2))^{1/n}.$$

It is easy to see that we cannot expect any inequality in the converse direction with a fixed constant and this even if we restrict ourselves to balls. However, Milman discovered that if A_1, A_2 are balls, there is always a *relative position* of A_1 and A_2 for which a converse inequality holds.

We will say that a ball $\widetilde{B}$ is *equivalent* to a ball B if there is a linear isomorphism $u : \mathbf{R}^n \to \mathbf{R}^n$ such that $|\det(u)| = 1$ and $\widetilde{B} = u(B)$.

Note that since u preserves volumes, we have $\operatorname{vol}(B) = \operatorname{vol}(\widetilde{B})$. Clearly, if D is a Milman ellipsoid for B, then $u(D)$ is a Milman ellipsoid for $u(B)$. Hence, for any ball B there is always an equivalent ball $\widetilde{B}$ admitting for its Milman's ellipsoid a multiple of the canonical ellipsoid $B_{\ell_2^n}$. *In that case we will say that $\widetilde{B}$ is in a regular position.* With this terminology, any ball is equivalent to a ball in a regular position, so that the next statement is very general.

Corollary 7.3. *Let $B_1, \ldots, B_k$ be balls in a regular position in $\mathbf{R}^n$. Then for all*

$$t_1 > 0, \ldots, t_k > 0$$

we have

(7.4)
$$
\begin{aligned}
(\operatorname{vol}(t_1 B_1 + \ldots &+ t_k B_k))^{1/n} \\
&\le C(3C)^k \left[t_1(\operatorname{vol}(B_1))^{1/n} + \ldots + t_k(\operatorname{vol}(B_k))^{1/n} \right]
\end{aligned}
$$

and

$$
\begin{aligned}
(\operatorname{vol}(t_1 B_1^\circ + \ldots &+ t_k B_k^\circ))^{1/n} \\
&\le C(3C)^k \left[t_1(\operatorname{vol}(B_1^\circ))^{1/n} + \ldots + t_k(\operatorname{vol}(B_k^\circ))^{1/n} \right],
\end{aligned}
$$

where C is as in Theorem 7.1.

Remark 7.4: Note that if $B_1, \ldots, B_k$ are in a regular position then the same is true for $t_1 B_1, \ldots t_k B_k$ and $t_1 B_1^\circ, \ldots, t_k B_k^\circ$. Therefore, it is enough to prove (7.4) and this only for $t_1 = \ldots = t_k = 1$.

To prove this corollary (and also in the sequel) we will need several elementary facts relating the covering (or entropy) numbers and volumes. Let A_1, A_2 be subsets of $\mathbf{R}^n$. Recall that we denote by $N(A_1, A_2)$ the smallest number N such that A_1 can be covered by N translates of A_2.

Clearly, we have

(7.5)
$$\operatorname{vol}(A_1) \le N(A_1, A_2) \operatorname{vol}(A_2).$$

More generally, we record here the following obvious facts.

For any subset $\Delta \subset \mathbf{R}^n$ (we also assume that all the sets appearing below are measurable) we have

$$(7.6) \qquad \mathrm{vol}(A_1 + \Delta) \leq N(A_1, A_2)\, \mathrm{vol}(A_2 + \Delta).$$

For arbitrary balls B_1, B_2 in $\mathbf{R}^n$ we have

$$(7.7) \qquad N(B_1, 2(B_1 \cap B_2)) \leq N(B_1, B_2).$$

Indeed, if $B_1 \subset \cup_{i \leq N}(x_i + B_2)$ then assuming $(x_i + B_1) \cap B_2 \neq \phi$, $\exists\, y_i \in B_1$ such that $x_i - y_i \in B_2$, hence $x_i + B_2 \subset y_i + 2B_2$ and

$$B_1 \cap (x_i + B_2) \subset B_1 \cap (y_i + 2B_2) \subset y_i + (B_1 - y_i) \cap (2B_2)$$
$$\subset y_i + 2(B_1 \cap B_2).$$

Therefore, $B_1 \subset \cup_{i \leq N}(y_i + 2(B_1 \cap B_2))$, so that (7.7) holds.

We will also use the following well-known fact (already proved above as Lemma 4.16).

Lemma 7.5.   *Let B_1, B_2 be balls in $\mathbf{R}^n$ with $B_2 \subset B_1$; then*

$$(7.8) \qquad \frac{\mathrm{vol}(B_1)}{\mathrm{vol}(B_2)} \leq N(B_1, B_2) \leq 3^n \frac{\mathrm{vol}(B_1)}{\mathrm{vol}(B_2)}.$$

Proof of Corollary 7.3: Let $D_1, \dots, D_k$ be Milman ellipsoids associated to $B_1, \dots, B_k$. By (7.8) and (7.2) we have $(i = 1, 2, \dots, k)$

$$N(B_i, D_i) \leq N(B_i + D_i, B_i \cap D_i) \leq 3^n C^n;$$

hence by (7.6) for any $\Delta \subset \mathbf{R}^n$

$$\mathrm{vol}(B_1 + \Delta)^{1/n} \leq 3C\, \mathrm{vol}(D_1 + \Delta)^{1/n}.$$

In particular,

$$\mathrm{vol}(B_1 + \dots + B_k)^{1/n} \leq 3C\, \mathrm{vol}(D_1 + B_2 + \dots + B_k)^{1/n}.$$

Repeating this argument for $B_2, \dots, B_k$, we find

$$\mathrm{vol}(B_1 + \dots + B_k)^{1/n} \leq (3C)^k\, \mathrm{vol}(D_1 + \dots + D_k)^{1/n}.$$

Now, since $B_1, \ldots, B_k$ are assumed in regular position, $D_1, \ldots, D_k$ are all multiples of a fixed ellipsoid (the Euclidean ball $B_{\ell_2^n}$), so that

$$\mathrm{vol}(D_1 + \ldots + D_k)^{1/n} = \mathrm{vol}(D_1)^{1/n} + \ldots + \mathrm{vol}(D_k)^{1/n},$$

and since

$$\left(\frac{\mathrm{vol}(D_i)}{\mathrm{vol}(B_i)}\right)^{1/n} \leq M(B_i, D_i) \leq C,$$

we obtain $\mathrm{vol}(B_1 + \ldots + B_k)^{1/n} \leq C(3C)^k[\mathrm{vol}(B_1)^{1/n} + \ldots + \mathrm{vol}(B_k)^{1/n}]$. Taking remark 7.4 into account, this completes the proof. ∎

Remark: Let us denote by $\mathrm{conv}(B \cup D)$ the convex hull of $B \cup D$. Then let

$$\widetilde{M}(B, D) = \left(\frac{\mathrm{vol}(\mathrm{conv}(B \cup D))}{\mathrm{vol}(B \cap D)} \cdot \frac{\mathrm{vol}(\mathrm{conv}(B^\circ \cup D^\circ))}{\mathrm{vol}(B^\circ \cap D^\circ)}\right)^{1/n}.$$

Clearly, Theorem 7.1 implies

$$\widetilde{M}(B, D) \leq C.$$

Although $M(B, D)$ is more flexible in the proof below, the ratio $\widetilde{M}(B, D)$ seems more natural, especially since $\widetilde{M}(B, D) = 1$ iff $B = D$.

The following lemma is the crucial tool for our first proof Theorem 7.1. The reader should note that only the second part of Lemma 7.6 is used below and to prove that part we can use Theorem 5.14 instead of Theorem 5.8.

Lemma 7.6. *There is a numerical constant K such that for all n, any n-dimensional Banach space E satisfies for all $v : E \to \ell_2^n$.*
(7.9)
$$\forall k \leq n \quad \max(e_k(v), e_k(v^*)) \leq K(1 + \mathrm{Log}\ vr(E))(n/k)^{3/2}\ell^*(v)n^{-1/2}.$$

In particular, for $k = n$,

$$\max(e_n(v), e_n(v^*)) \leq K(1 + \mathrm{Log}\ vr(E))\ell^*(v)n^{-1/2}.$$

Proof: Let $A = vr(E)$.

By Theorem 6.1, for every $k \leq n$, there is a subspace $F \subset E$ such that $d_F \leq (4\pi A)^{n/k}$ and codim $F = k$. Consider the restriction $v_{|F}$. We have, by Lemma 3.10.,

$$\ell((v_{|F})^*) \leq K(F)\ell^*(v_{|F}) \leq K(F)\ell^*(v).$$

By Theorem 2.5, $K(F) \leq K(1 + \text{Log } d_F)$, hence

$$K(F) \leq K(1 + \frac{n}{k}\text{Log}(4\pi A)).$$

By Theorem 5.8 $k^{1/2}c_k(v_{|F}) \leq c'\ell((v_{|F})^*)$, hence we have

$$c_{2k}(v) \leq c_k(v_{|F}) \leq 2Kc' \left(\frac{n}{k}\right)^{3/2} \text{Log}(4\pi A)\ell^*(v)n^{-1/2}.$$

This implies that for some constant c'' we have

$$\sup_{k \leq n} k^{3/2}c_k(v) \leq n^{3/2}c''(1 + \text{Log } A)\ell^*(v)n^{-1/2},$$

so that (7.9) follows immediately, by Theorem 5.2. ∎

Now we can easily deduce from Lemma 7.6 the following preliminary form of Theorem 7.1.

Proposition 7.7.   *There is a numerical constant K_1 such that for all $n \geq 1$, for every n-dimensional normed space E with unit ball $B_E \subset \mathbf{R}^n$, there is an ellipsoid $D_1 \subset \mathbf{R}^n$ such that*

$$M(B_E, D_1) \leq K_1(1 + \text{Log } vr(E))^2.$$

Proof: We consider the Lewis ellipsoid associated to E as in Theorem 3.1, i.e. we have $u : \ell_2^n \to E$ satisfying $\ell(u) = n^{1/2}$ and $\ell^*(u^{-1}) = n^{1/2}$. By Lemma 7.6, we have

(7.10) $\max(e_n(u^{-1}), e_n(u^{-1*})) \leq K(1 + \text{Log } vr(E)).$

On the other hand, by Corollary 5.12, we have

(7.11) $\max(e_n(u), e_n(u^*)) \leq K'$

for some numerical constant K' ($K' = c_3(c_4)^{-1}$).

Let us denote simply $B = B_E$ and $D_1 = u(B_{\ell_2^n})$.

Let $\lambda = K(1 + \text{Log } vr(E))$. We have by (7.10) and (7.11)

(7.12)
$$N(B, \lambda D_1) \leq 2^{n-1}, N(D_1^\circ, \lambda B^\circ) \leq 2^{n-1},$$
$$N(D_1, K'B) \leq 2^{n-1}, N(B^\circ, K'D_1^\circ) \leq 2^{n-1}.$$

By (7.6), this implies

$$(7.13) \quad \begin{aligned} \mathrm{vol}(B + D_1) &\leq (1 + \lambda)^n 2^{n-1} \mathrm{vol}(D_1), \\ \mathrm{vol}(B^\circ + D_1^\circ) &\leq (1 + K')^n 2^{n-1} \mathrm{vol}(D_1^\circ). \end{aligned}$$

By (7.5) we have

$$\mathrm{vol}(D_1) \leq N(D_1, 2(D_1 \cap K'B)) \, \mathrm{vol}(2(D_1 \cap K'B));$$

hence, by (7.7) and (7.12),

$$\mathrm{vol}(D_1) \leq 2^{n-1}(2 + 2K')^n \, \mathrm{vol}(D_1 \cap B).$$

Similarly, we find

$$\mathrm{vol}(D_1^\circ) \leq 2^{n-1}(2 + 2\lambda)^n \, \mathrm{vol}(D_1^\circ \cap B^\circ).$$

Therefore, by (7.13) and the last two inequalities, we have

$$M(B, D_1) \leq 16(1 + \lambda)^2 (1 + K')^2,$$

and Proposition 7.7 follows immediately

First Proof of Theorem 7.1: Let $n \geq 1$ be a fixed integer. Let C_n be the smallest constant C for which the statement of Theorem 7.1 is true, that is to say

$$C_n = \sup_{\substack{B \subset \mathbf{R}^n \\ B \text{ ball}}} \inf_{\substack{D \subset \mathbf{R}^n \\ D \text{ ellipsoid}}} M(B, D).$$

We will show by an *a priori* estimate that C_n has to remain bounded when $n \to \infty$. (Of course, we know that C_n is finite since for any B there is an ellipsoid D—the John ellipsoid—such that $D \subset B \subset \sqrt{n}D$, which implies $C_n \leq 2n^{1/2}(n^{1/2} + 1)$.)

Let D be an ellipsoid such that

$$(7.14) \quad M(B, D) \leq 4C_n.$$

Clearly, we have $M(B, D) = M_1 M_2$ where

$$M_1 = \left(\frac{\mathrm{vol}(B + D)}{\mathrm{vol}(B)} \cdot \frac{\mathrm{vol}(B^\circ)}{\mathrm{vol}(B^\circ \cap D^\circ)} \right)^{1/n}$$

and

$$M_2 = \left(\frac{\text{vol}(B^\circ + D^\circ)}{\text{vol}(B^\circ)} \cdot \frac{\text{vol}(B)}{\text{vol}(B \cap D)} \right)^{1/n}.$$

Note that M_2 is obtained from M_1 by replacing B, D by the polars B°, D°.

By (7.14) we have either $M_1 \le 2(C_n)^{1/2}$ or $M_2 \le 2(C_n)^{1/2}$. We will show that $M_1 \le 2(C_n)^{1/2}$ implies that there is an ellipsoid $D_1 \subset \mathbf{R}^n$ satisfying

$$M(B, D_1) \le K_2 (C_n)^{1/2} (1 + \text{Log } C_n)^2$$

for some numerical constant K_2.

Since $M(B, D_1) = M(B^\circ, D_1^\circ)$, the other case $M_2 \le 2(C_n)^{1/2}$ leads to the same conclusion by simply exchanging the roles of (B, D) and (B°, D°).

Hence we assume $M_1 \le 2(C_n)^{1/2}$.

Let

$$\alpha = \left(\frac{\text{vol}(B + D)}{\text{vol}(B)} \right)^{1/n} \quad \text{and} \quad \beta = \left(\frac{\text{vol}(B^\circ)}{\text{vol}(B^\circ \cap D^\circ)} \right)^{1/n}$$

so that $M_1 = \alpha\beta \le 2(C_n)^{1/2}$.

By Lemma 7.5 we have

(7.15) $$N(B + D, B) \le (3\alpha)^n,$$

(7.16) $$N(B^\circ, B^\circ \cap D^\circ) \le (3\beta)^n.$$

Let us denote by E_+ the space $\mathbf{R}^n$ equipped with the norm admitting $B + D$ as its unit ball.

Then, clearly,

$$vr(E_+) \le \left(\frac{\text{vol}(B + D)}{\text{vol}(D)} \right)^{1/n} \le M(B, D) \le 4C_n.$$

We now apply Proposition 7.7 to the space E_+. This implies the existence of an ellipsoid D_1 such that

$$M(B + D, D_1) \le K_1 (1 + \text{Log}(4C_n))^2.$$

But we may use (7.15) and (7.16) to majorize $M(B, D_1)$ by a suitable multiple of $M(B + D, D_1)$. More precisely, here are the details.

First we have trivially

$$(7.17) \qquad \mathrm{vol}(B + D_1) \leq \mathrm{vol}(B + D + D_1),$$

and $N((B + D) \cap D_1, B) \leq N(B + D, B)$ implies, by (7.15) and (7.7),

$$N((B + D) \cap D_1, 2(B \cap D_1)) \leq (3\alpha)^n.$$

Therefore,

$$(7.18) \qquad \mathrm{vol}((B + D) \cap D_1) \leq (6\alpha)^n \, \mathrm{vol}(B \cap D_1).$$

Putting (7.17) and (7.18) together we find

$$(7.19) \qquad \frac{\mathrm{vol}(B + D_1)}{\mathrm{vol}(B \cap D_1)} \leq \frac{\mathrm{vol}(B + D + D_1)}{\mathrm{vol}((B + D) \cap D_1)} (6\alpha)^n.$$

On the other hand, we have trivially

$$(7.20) \qquad \mathrm{vol}((B + D)^\circ \cap D_1^\circ) \leq \mathrm{vol}(B^\circ \cap D_1^\circ),$$

and (7.16) implies (with (7.6))

$$(7.21) \qquad \begin{aligned} \mathrm{vol}(B^\circ + D_1^\circ) &\leq (3\beta)^n \, \mathrm{vol}(B^\circ \cap D^\circ + D_1^\circ) \\ &\leq (6\beta)^n \, \mathrm{vol}((B + D)^\circ + D_1^\circ). \end{aligned}$$

Now, (7.20) and (7.21) yield

$$(7.22) \qquad \frac{\mathrm{vol}(B^\circ + D_1^\circ)}{\mathrm{vol}(B^\circ \cap D_1^\circ)} \leq \frac{\mathrm{vol}((B + D)^\circ + D_1^\circ)}{\mathrm{vol}((B + D)^\circ \cap D_1^\circ)} (6\beta)^n.$$

Putting (7.19) and (7.22) together yields

$$M(B, D_1) \leq 36\alpha\beta \cdot M(B + D, D_1);$$

hence (recalling $\alpha\beta = M_1 \leq 2(C_n)^{1/2}$)

$$\leq 72(C_n)^{1/2} K_1 (1 + \mathrm{Log}(4C_n))^2.$$

In conclusion, we obtain as announced

$$C_n \leq 72(C_n)^{1/2} K_1 (1 + \mathrm{Log}(4C_n))^2,$$

so that C_n must remain bounded when $n \to \infty$, hence $C = \sup_{n \geq 1} C_n$ is finite. ∎

We now give more corollaries of Theorem 7.1.

Corollary 7.8. *Let B_1, B_2 be balls in $\mathbf{R}^n$ in a regular position. For some numerical constant $\overline{K}$ ($\overline{K} \leq 2(3C)^3$) we have for all $\varepsilon > 0$*
(7.23)
$$\max\left\{1, \varepsilon^{-n}\frac{\text{vol}(B_1)}{\text{vol}(B_2)}\right\} \leq N(B_1, \varepsilon B_2) \leq (\overline{K})^n \max\left\{1, \varepsilon^{-n}\frac{\text{vol}(B_1)}{\text{vol}(B_2)}\right\}.$$

Proof: The left side of (7.23) is trivial.
By Corollary 7.3, we have

$$\frac{\text{vol}(B_1 + \varepsilon B_2)}{\varepsilon\,\text{vol}(B_2)} \leq (C(3C)^2)^n \left(1 + \varepsilon^{-n}\frac{\text{vol}(B_1)}{\text{vol}(B_2)}\right).$$

Hence the right side of (7.23) follows from lemma 7.5.

We can also deduce the *quotient of a subspace theorem* (for short the QS Theorem) of Milman from Theorem 7.1, using the original approach of the paper [M3].

Corollary 7.9. *There is a function $f :]0,1[\to \mathbf{R}_+$ such that, for any $n \geq 1$, for any $0 < \delta < 1$, every n-dimensional normed space E contains subspaces $F_2 \subset E_1$ with $\dim(E_1/F_2) = [(1 - \delta)n]$ and such that $d_{E_1/F_2} \leq f(\delta)$.*

Note: Of course $f(\delta) \to \infty$ when $\delta \to 0$. We will get a more precise estimate of $f(\delta)$ in the next chapter, so we do not insist on the behavior here.

Proof: Let $B \subset \mathbf{R}^n$ be the unit ball of E and let D be an associated Milman ellipsoid such that $M(B, D) \leq C$. By (7.1) we have

$$\left(\frac{\text{vol}(B + D)}{\text{vol}(D)}\right)^{1/n} \leq C.$$

Hence, by Theorem 6.1, for any $k = 1, \dots, n-1$ there is a subspace $E_1 \subset \mathbf{R}^n$ with $\dim E_1 = k$ such that

$$(B + D) \cap E_1 \subset (4\pi C)^{\frac{n}{n-k}} D \cap E_1.$$

A fortiori we have $B \cap E_1 \subset (4\pi C)^{\frac{n}{n-k}} D \cap E_1$.
Let $\lambda = (4\pi C)^{\frac{n}{n-k}}$. Taking polars, we have

(7.24) $P_{E_1}(B^\circ) \supset \lambda^{-1} P_{E_1}(D^\circ).$

Assume $k \geq \frac{n}{2}$. We then claim that

$$(7.25) \qquad \left(\frac{\text{vol}(P_{E_1}(B^\circ))}{\text{vol}(P_{E_1}(D^\circ))} \right)^{1/k} \leq (2C)^2.$$

Indeed, we have by Lemma 8.8 below

$$\text{vol}(P_{E_1}(B^\circ)) \leq 2^n \frac{\text{vol}(B^\circ)}{\text{vol}(E_1^\perp \cap B^\circ)} \leq 2^n C^n \frac{\text{vol}(B^\circ \cap D^\circ)}{\text{vol}(E_1^\perp \cap B^\circ)}$$

$$\leq 2^n C^n \frac{\text{vol}(P_{E_1}(B^\circ \cap D^\circ)) \, \text{vol}(E_1^\perp \cap B^\circ \cap D^\circ)}{\text{vol}(E_1^\perp \cap B^\circ)}$$

$$\leq 2^n C^n \, \text{vol}(P_{E_1}(D^\circ)).$$

This proves the above claim.

We now go back to (7.24). This implies with (7.25) that $vr(P_{E_1}(B^\circ))$ $\leq (2C)^2 \lambda$. Therefore, applying Theorem 6.1 again we find a subspace $E_2 \subset E_1$ with $\dim E_2 = \ell$ such that if we denote $D_1^\circ = \lambda^{-1} P_{E_1}(D^\circ)$ we have

$$D_1^\circ \cap E_2 \subset P_{E_1}(B^\circ) \cap E_2 \subset (4\pi(2C)^2\lambda)^{k/(k-\ell)} D_1^\circ \cap E_2.$$

Taking polars again,

$$P_{E_2}(D_1) \supset P_{E_2}(E_1 \cap B) \supset (4\pi(2C)^2\lambda)^{-k/(k-\ell)} P_{E_2}(D_1);$$

hence, since $P_{E_2}(D_1)$ is an ellipsoid and (cf. the beginning of Chapter 1), since we may identify $P_{E_2}(E_1 \cap B)$ with the unit ball of E_1/F_2 for $F_2 = E_1 \cap E_2^\perp$, we have

$$d_{E_1/F_2} \leq (4\pi(2C)^2\lambda)^{k/(k-\ell)}.$$

Finally, choose k, ℓ such that $n - k = \left[\frac{1-\delta}{2}n \right]$ and $k - \ell = \left[\frac{1-\delta}{2} \cdot n \right]$; then $n - \ell \leq (1-\delta)n$, so that $\ell \geq \delta n$ and for some numerical constant C' we have

$$d_{E_1/F_2} \leq C' \left(4\pi(2C)^2(4\pi C)^{2/(1-\delta)} \right)^{2/(1-\delta)}. \qquad \blacksquare$$

Corollary 7.10. *Let $B_1, \dots, B_k$ be balls in a regular position in $\mathbf{R}^n$. Then there is a constant $C(k)$ depending only on k such that for all $t_1 > 0, \dots, t_k > 0$ we have*

$$\inf \left\{ t_1 \, \text{vol}(B_1)^{1/n}, \dots, t_k \, \text{vol}(B_k)^{1/n} \right\} \leq C(k) \, \text{vol}(t_1 B_1 \cap \dots \cap t_k B_k)^{1/n}.$$

Proof: The proof is an easy consequence of Corollaries 7.3 and 7.2 and is left as an exercise for the reader. [Note that this reduces to the case $t_1 = \ldots = t_k = 1$ and that by Corollary 7.2 $\mathrm{vol}(B_1 \cap B_2)^{1/n}$ is essentially equivalent to $n^{-1} \mathrm{vol}(B_1^\circ + B_2^\circ)^{1/n}$.] ∎

Recently, we found a different approach to the results of this chapter. The basic ingredients such as Lemma 3.10 and either Theorem 5.8 or Theorem 5.14 are the same, but we use an interpolation method (the real or the complex method) instead of Lemma 7.6 and Proposition 7.7.

As the reader has observed, in the first proof of Theorem 7.1 the main point is the following: Given our ball B and any ellipsoid D, we can apply Lemma 3.10 either to $B + D$ or to $B \cap D$ (depending on which is closest to B); this yields a new ellipsoid D_1 for which $M(B, D_1)$ is significantly smaller than $M(B, D)$. This shows that the best constant C in Theorem 7.1 must remain bounded. (Alternatively, we could repeat the preceding procedure, to construct a sequence of ellipsoids $D_1, D_2, \ldots, D_k, \ldots$; then any limit point of this sequence yields an ellipsoid satisfying the conclusion of Theorem 7.1.)

The second proof is similar in spirit but instead of considering $B + D$ and $B \cap D$, we consider an *interpolation space* between the normed spaces associated to B and D. Roughly speaking, we introduce a family of "interpolated balls" $B_\theta (0 < \theta < 1)$ such that $B_0 = B$, $B_1 = D$ which constitute a continuous deformation of B into D possessing the crucial property that the associated normed space E_θ admits a K-convexity constant majorized by a function of θ alone (i.e. we have $K(E_\theta) \leq \phi(\theta)$ with $\phi(\theta)$ depending only on θ).

To make the construction of E_θ more precise, we need the definition and the basic properties of the complex interpolation spaces, originally introduced by A. Calderón and J. L. Lions (cf. [BL] or [KPS]).

Let E_0, E_1 be a couple of Banach spaces. We will assume that they are "compatible" in the following sense: E_0 and E_1 are both continuously embedded into a larger topological vector space, so that we may consider unambiguously the spaces $E_0 + E_1$ and $E_0 \cap E_1$. Actually, in the construction below, E_0 and E_1 will be simply the same space $\mathbf{C}^n$ equipped with two different norms. Therefore we do not insist on several technicalities which are irrelevant for our present purposes.

Let E_0, E_1 be two *complex* Banach spaces, compatible in the above sense. To define the complex interpolation spaces between E_0 and E_1, we need some notation. Let

$$S = \{z \in \mathbf{C} | 0 < Re\, z < 1\}$$
$$\overline{S} = \{z \in \mathbf{C} | 0 \leq Re\, z \leq 1\}$$
$$S_0 = \{z \in \mathbf{C} | Re\, z = 0\}$$
$$S_1 = \{z \in \mathbf{C} | Re\, z = 1\}.$$

Let $\mathcal{F}$ be the space of all bounded continuous functions $f : \overline{S} \to E_0 + E_1$ which are holomorphic on the strip S and such that the restriction $f_{|S_0}$ assumes values in the space E_0 and is continuous and bounded for the norm of E_0, while the restriction $f_{|S_1}$ assumes values in the space E_1 and is continuous and bounded for the norm of E_1.

We equip this space with the norm

$$\|f\|_{\mathcal{F}} = \max \left\{ \sup_{t\in\mathbf{R}} \|f(it)\|_{E_0}, \sup_{t\in\mathbf{R}} \|f(1+it)\|_{E_1} \right\}.$$

We now define the interpolation space $(E_0, E_1)_\theta$, which we will denote simply by E_θ, as the set of all x in $E_0 + E_1$ for which there is an f in $\mathcal{F}$ such that $x = f(\theta)$. We equip this space with the norm

$$\|x\|_\theta = \inf \{\|f\|_{\mathcal{F}} | f \in \mathcal{F} \quad f(\theta) = x\}.$$

It is rather simple to check that this defines a new Banach space continuously embedded into $E_0 + E_1$. The fundamental property of these spaces is the "interpolation theorem" which states that if (F_0, F_1) is another similar couple, and if an operator T maps boundedly both E_0 into F_0 and E_1 into F_1, then T maps boundedly E_θ into F_θ and we have

$$\|T\|_{E_\theta \to F_\theta} \leq (\|T\|_{E_0 \to F_0})^{1-\theta} (\|T\|_{E_1 \to F_1})^\theta.$$

In the application below, we need the following simpler form of this general principle.

$$(7.26) \qquad \forall\, x \in E_\theta \quad \|x\|_{E_\theta} \leq \|x\|_{E_0}^{1-\theta} \|x\|_{E_1}^\theta.$$

This is very easy to check. Let $M_0 = \|x\|_{E_0}, M_1 = \|x\|_{E_1}$ and let $f(z) = M_0^{1-\theta} M_1^\theta (M_0^{1-z} M_1^z)^{-1} x$. Clearly, $f \in \mathcal{F}, \|f\|_{\mathcal{F}} = M_0^{1-\theta} M_1^\theta$ and $f(\theta) = x$, so that (7.26) follows immediately.

We also need the dual inequality for which we assume that $E_0 \cap E_1$ is dense both in E_0 and in E_1. For all linear forms ζ in $(E_0 \cap E_1)^*$ we denote if $0 \leq \theta \leq 1$

$$\|\zeta\|_\theta^* = \sup \{|\zeta(x)| \quad x \in E_0 \cap E_1 \quad \|x\|_{E_\theta} \leq 1\}.$$

Then we have

$(7.26)^*$ $\forall \, \zeta \in (E_0 \cap E_1)^*$ $\|\zeta\|_\theta^* \leq (\|\zeta\|_0^*)^{1-\theta} \, (\|\zeta\|_1^*)^\theta .$

This is a simple consequence of the "three-lines lemma", a classical result which says that any function $g : \overline{S} \to \mathbf{C}$ which is bounded, continuous and holomorphic on S must satisfy (cf. e.g. [BL])

$$|g(\theta)| \leq \left(\max_{S_0} |g| \right)^{1-\theta} \left(\max_{S_1} |g| \right)^\theta .$$

Indeed, consider ζ in $(E_0 \cap E_1)^*$; let x in $E_0 \cap E_1$ be such that $\|x\|_\theta < 1$. Then choose f in $\mathcal{F}$ such that $\|f\|_{\mathcal{F}} < 1$ and $f(\theta) = x$.

We may apply the three-lines lemma to

$$g(z) = \langle \zeta, f(z) \rangle.$$

This yields

$$|\langle \zeta, x \rangle| = |g(\theta)| \leq (\|\zeta\|_0^*)^{1-\theta} \, (\|\zeta\|_1^*)^\theta$$

and this establishes $(7.26)^*$.

Now let E be an other complex Banach space, and let $T : E_0 + E_1 \to E$ and $R : E \to E_0 \cap E_1$ be bounded linear (meaning, of course, **C**-linear) operators. *A fortiori*, T is bounded from E_θ into E and R is bounded from E into E_θ. We will denote by $T_\theta : E_\theta \to E$ and by $R_\theta : E \to E_\theta$ the corresponding operators.

We will use the following elementary facts.

(7.27) $c_k(R_\theta) \leq \|R_0\|^{1-\theta} \, (c_k(R_1))^\theta$

(7.28) $d_k(T_\theta) \leq \|T_0\|^{1-\theta} \, (d_k(T_1))^\theta .$

The inequality (7.27) is an immediate consequence of (7.26). For simplicity, we assume in (7.28) that T_θ and T_1 are compact operators and that $E_0 \cap E_1$ is dense both in E_0 and E_1. These assumptions are trivially verified in the application below. Then, since $d_k(T_\theta) = c_k(T_\theta^*)$ and $d_k(T_1) = c_k(T_1^*)$, the inequality (7.28) appears as an immediate consequence of $(7.26)^*$.

We come finally to the result which we use as a substitute for the logarithmic bound of the K-convexity constant (cf. Theorem 2.5).

Theorem 7.11. *Let (E_0, E_1) be a compatible couple of complex Banach spaces as above. If E_1 is a Hilbert space then for all $0 < \theta < 1$ the space E_θ is K-convex, and we have*

$$K(E_\theta) \le \phi(\theta)$$

for some constant $\phi(\theta)$ depending only on θ. Moreover, the function $\theta \to \phi(\theta)$ is $0(1/\theta)$ when $\theta \to 0$.

We will use the following elementary fact.

Lemma 7.12. *For $z \in \mathbf{C}$, we denote by $\operatorname{Arg}(z)$ the argument of z determined so that $-\pi < \operatorname{Arg}(z) \le \pi$. Let $D_0 = [-1, 1]$, $D_1 = \{z \in \mathbf{C} \mid |z| = 1\}$ and for $0 < \theta < 1$ let*

$$D_\theta = \left\{ z \in \mathbf{C} \,\middle|\, \left| \operatorname{Arg}\left(\frac{z+1}{z-1}\right) \right| = \theta\pi/2 \right\}.$$

Then for all ζ in D_θ there is a continuous function $\Delta : \overline{S} \to \mathbf{C}$, holomorphic on S and such that

(7.29)
$$\Delta(\zeta) \in D_0 \quad \forall\, \zeta \in S_0,$$

(7.30)
$$\Delta(\zeta) \in D_1 \quad \forall\, \zeta \in S_1,$$

and
$$\Delta(\theta) = \zeta.$$

Proof: We remark that D_θ is the set of points m in the complex plane from which the segment $[-1, 1]$ lies in an angle exactly equal to $\pi - \pi\theta/2$. (See the picture, the angle with vertex m in the triangle formed with m by -1 and $+1$ is equal to $\pi - \pi\theta/2$.)

Now pick a number ζ in D_θ. Assume $\operatorname{Im} \zeta > 0$ (resp. $\operatorname{Im} \zeta < 0$). Then it is easy to check that there is a real number t such that $\zeta = \mathrm{i}\tan\left(\frac{\pi}{4}\theta + \mathrm{i}t\right)$ (resp. $\zeta = -\mathrm{i}\tan\left(\frac{\pi}{4}\theta + \mathrm{i}t\right)$). We can then define

$$\Delta(z) = \mathrm{i}\tan\left(\frac{\pi}{4}z + \mathrm{i}t\right)$$

(resp. $-\mathrm{i}\tan\left(\frac{\pi}{4}z + \mathrm{i}t\right)$).

Clearly Δ is holomorphic on a neighbourhood of S and satisfies the conclusions of lemma 7.12.

(**Note:** The key point implicit in this fact is a property of the Harmonic measure for the domain bounded by the segment $[-1,1]$ and the upper half of D_1. The function $z \to \frac{2}{\pi} \operatorname{Arg}\left(\frac{z+1}{z-1}\right)$ is harmonic on this domain and takes the value 1 on the upper half of D_1 and 0 on D_0.)

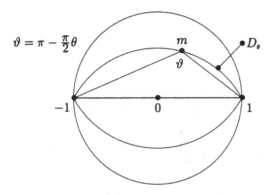

Proof of Theorem 7.11: We use the same notation as in Chapter 2 (cf. Lemma 2.1). For any z in $\mathbf{C}$ with $|z| \le 1$, we define

$$T(z) = \sum_{k \ge 0} z^k Q_k.$$

This defines an operator of norm 1 on $L_2 = L_2(\Omega, \mathcal{A}, \mathbf{P})$. Fix a number $0 < \theta < 1$. We will find a number $\delta = \delta(\theta) > 0$ such that for all ζ with $|\zeta| \le \delta$ we have

$$\|T(z) \otimes I_{E_\theta}\|_{L_2(E_\theta) \to L_2(E_\theta)} \le 1.$$

Let $\widetilde{T}(z) = T(z) \otimes I_{E_\theta}$ and $\widetilde{Q}_1 = Q_1 \otimes I_{E_\theta}$. By Cauchy's formula, it is easy to check that

$$Q_1 = \frac{1}{\delta} \int_0^{2\pi} e^{-is} T(\delta e^{is}) \frac{ds}{2\pi},$$

and of course the same identity holds between $\widetilde{Q}_1$ and $\widetilde{T}(z)$. Therefore, by Jensen's inequality

(7.31)
$$K(E_\theta) = \left\|\widetilde{Q}_1\right\|_{L_2(E_\theta) \to L_2(E_\theta)} \leq \delta^{-1} \sup_{|\zeta|=\delta} \left\|\widetilde{T}(\zeta)\right\|_{L_2(E_\theta) \to L_2(E_\theta)}.$$

We thus obtain the desired conclusion provided a suitable $\delta = \delta(\theta)$ has been found. We now explain how to find δ.

Recall that for an arbitrary Banach space E_0 we have by Lemma 2.2

(7.32)
$$\forall \, \zeta \in D_0 \quad \|T(z) \otimes I_{E_0}\|_{L_2(E_0) \to L_2(E_0)} \leq 1;$$

on the other hand, if E_1 is a Hilbert space we have (again by Lemma 2.2)

(7.33)
$$\forall \, z \in D_1 \quad \|T(z) \otimes I_{E_1}\|_{L_2(E_1) \to L_2(E_1)} \leq 1.$$

We can then use Lemma 7.12 to "interpolate" between (7.32) and (7.33) and we obtain that for all ζ in D_θ the operator $T(\zeta) \otimes I_{E_\theta}$ is bounded on $L_2(E_\theta)$ with norm ≤ 1. Let us quickly sketch the proof of this.

First we use the well-known fact that (cf. [BL] p. 107).

(7.34)
$$(L_2(E_0), L_2(E_1))_\theta = L_2(E_\theta)$$

isometrically. Then for any ϕ in $L_2(E_\theta)$ with norm < 1, there is a function f in the space $\mathcal{F}$ associated to the couple $(L_2(E_0), L_2(E_1))$ with $\|f\|_{\mathcal{F}} < 1$ and $f(\theta) = \phi$. Fix ζ in D_θ and let Δ be as in Lemma 7.12 so that in particular $\Delta(\theta) = \zeta$. We can introduce the function $G(z) = T(\Delta(z))f(z)$ with values in $L_2(E_0) + L_2(E_1)$. This function is holomorphic in S and $G(\theta) = T(\zeta)\phi$.

Let us denote $\wedge_0 = L_2(E_0), \wedge_1 = L_2(E_1)$. Clearly, we have

$$\|T(\zeta)\phi\|_{(\wedge_0,\wedge_1)_\theta} \leq \max \left\{ \|G\|_{L^\infty(S_0,\wedge_0)}, \|G\|_{L^\infty(S_1,\wedge_1)} \right\};$$

hence, by (7.32) and (7.33) (recalling (7.29) and (7.30)),

$$\|T(\zeta)\phi\|_{(\wedge_0,\wedge_1)_\theta} \leq 1.$$

Therefore by (7.34) we conclude that

(7.35)
$$\forall \, \zeta \in D_\theta \quad \|T(\zeta)\|_{L_2(E_\theta) \to L_2(E_\theta)} \leq 1.$$

Clearly (since $T(z_1 z_2) = T(z_1)T(z_2)$ and by (7.32)), the upper bound (7.35) remains valid for any ζ inside the domain circled by D_θ. In particular, we have (7.35) for all ζ in $\mathbf{C}$ such that $|\zeta| \leq \tan(\theta\pi/4)$. By (7.31), we finally conclude

$$K(E_\theta) \leq (\tan(\theta\pi/4))^{-1}. \quad \blacksquare$$

Remark: It seems worthwhile to observe that Theorem 2.5 is a corollary of Theorem 7.11. Indeed, we may assume in Theorem 2.5 that X is a finite-dimensional complex space. We may as well assume that $X = \mathbf{C}^n$ and that the operator $u : \ell_2^n \to X$ which satisfies $\|u\| = d(X, \ell_2^n)$ and $\|u^{-1}\| = 1$ is nothing but the identity operator. Then let E_θ be the interpolation space associated to the couple $E_0 = X, E_1 = \ell_2^n$. By the interpolation theorem (or by (7.27) and (7.28) for $k = 1$) we have

$$d(E, E_\theta) \leq d(X, \ell_2^n)^\theta$$

and by Theorem 7.11 we have $K(E_\theta) \leq \phi(\theta)$; hence

$$K(E) \leq d(E, E_\theta)K(E_\theta) \leq \phi(\theta)d(X, \ell_2^n)^\theta$$
$$\leq (K^{'}/\theta)d(X, \ell_2^n)^\theta$$

for some numerical constant $K^{'}$. It remains to optimize the choice of $0 < \theta < 1$ and we obtain

$$K(E) \leq K(1 + \log d(X, \ell_2^n)),$$

as in Theorem 2.5. $\quad \blacksquare$

We now present a different approach to Theorem 7.1. Actually, we will prove the following refinement of Theorem 7.1 (similar in spirit to Theorem 3.1).

Theorem 7.13. *For each $\alpha > 1/2$, there is a constant $C = C(\alpha)$ such that for any n and any n-dimensional (real or complex) normed space E, there is an isomorphism $u : \ell_2^n \to E$ such that*

(7.36) $\forall k = 1, 2, \ldots n \quad d_k(u) \leq C(n/k)^\alpha \text{ and } c_k(u^{-1}) \leq C(n/k)^\alpha.$

Moreover, the constant $C(\alpha)$ is $0((\alpha - 2^{-1})^{-1/2})$ when $\alpha \to 1/2$.

Note: At this point the reader should recall that $d_k(u) = c_k(u^*)$.

Remark 7.14: Like most of the preceding results, Theorem 7.13 is immediate if the K-convexity constant of E is bounded above. Indeed, by Theorem 3.11 there is an isomorphism $v : \ell_2^n \to E$ such that

$$\ell(v) \leq n^{1/2} \quad \text{and} \quad \ell(v^{-1*}) \leq K(E)n^{1/2};$$

hence by Theorem 5.8 we have

$$d_k(v) \leq C'(n/k)^{1/2} \qquad c_k(v^{-1}) \leq K(E)C'(n/k)^{1/2}.$$

We note in passing that although we have proved this in the real case only, the results hold in the complex case also with identical proofs.

Proof of Theorem 7.13: We first consider the complex case. We denote again by ℓ_2^n the complex n-dimensional Euclidean space. Let us fix $\alpha > 1/2$ and consider an n-dimensional normed space E over the complex field. We denote *a priori* by C the best constant for which the conclusion of Theorem 7.13 is valid. To complete the proof, we have to show that C is bounded above by a function of α alone. By definition of C, there is a **C**-linear isomorphism $u : \ell_2^n \to E$ such that (7.36) holds. Without loss of generality we may assume that $\mathbf{C}^n$ is the underlying vector space of E and that u is the identity operator.

Let us define $0 < \theta < 1$ such that $\theta = \alpha^{-1}(\alpha - 2^{-1})$. We will consider the interpolation space $E_\theta = (E, \ell_2^n)_\theta$ associated to the couple $E_0 = E, E_1 = \ell_2^n$. As before, we denote by $u_\theta : E_\theta \to E$ the identity operator considered as acting from E_θ into E. By Theorem 7.11, the space E_θ can be viewed as a K-convex perturbation of E, to which Remark 7.14 can be applied. Therefore, there is an isomorphism $v : \ell_2^n \to E_\theta$ such that
(7.37)
$$\forall\, k = 1, \dots n \quad d_k(v) \leq C'(n/k)^{1/2} \quad \text{and} \quad c_k(v^{-1}) \leq C'\phi(\theta)(n/k)^{1/2}.$$

If we wish to replace E_θ by E in (7.37), we need to evaluate $d_k(u_\theta)$ and $c_k(u_\theta^{-1})$. This is easy using (7.27) and (7.28). Indeed since $u_0 = I_E : E \to E$ and $u_1 = u : \ell_2^n \to E$, (7.27) and (7.28) yield

$$d_k(u_\theta) \leq \|u_0\|^{1-\theta}d_k(u_1)^\theta = d_k(u)^\theta$$

and

$$c_k(u_\theta^{-1}) \leq \|u_0^{-1}\|^{1-\theta}\, c_k(u_1^{-1})^\theta = c_k(u^{-1})^\theta.$$

Hence, recalling our *a priori* assumption that u satisfies (7.36), we have
(7.38)
$$\forall\, k \geq 1 \quad d_k(u_\theta) \leq (C(n/k)^\alpha)^\theta \quad \text{and} \quad c_k(u_\theta^{-1}) \leq (C(n/k)^\alpha)^\theta.$$

Let us now evaluate the operator $w = u_\theta v : \ell_2^n \to E$ in connection with the conclusions of Theorem 7.13.

We have, by (5.1),

$$d_{2k-1}(w) \le d_k(u_\theta) d_k(v);$$

hence, by (7.37) and (7.38)

$$\le C' C^\theta (n/k)^{\alpha\theta+1/2} = C' C^\theta (n/k)^\alpha.$$

Similarly,

$$c_{2k-1}(w^{-1}) \le c_k(u_\theta^{-1}) c_k(v^{-1}) \le C' \phi(\theta) C^\theta (n/k)^\alpha.$$

Let $\lambda = 2^\alpha C' C^\theta$. By an elementary computation, these bounds imply

$$\forall\, k \ge 1 \quad d_k(w) \le \lambda(n/k)^\alpha \ \text{ and } \ c_k(w^{-1}) \le \lambda\phi(\theta)(n/k)^\alpha,$$

and if we replace w by $\widetilde{w} = \phi(\theta)^{1/2} w$, we find
(7.39)
$$\forall\, k \ge 1 \quad d_k(\widetilde{w}) \le \lambda\phi(\theta)^{1/2}(n/k)^\alpha \ \text{ and } \ c_k(\widetilde{w}^{-1}) \le \lambda\phi(\theta)^{1/2}(n/k)^\alpha.$$

Returning to the *a priori* definition of C as the smallest possible constant in Theorem 7.13, we see that (7.39) implies

$$C \le \lambda\phi(\theta)^{1/2} = C' C^\theta 2^\alpha \phi(\theta)^{1/2};$$

hence dividing by C^θ we finally obtain $C \le C(\alpha)$ with

$$C(\alpha) = \left(C' 2^\alpha \phi(\theta)^{1/2} \right)^{\frac{1}{1-\theta}} \ \text{ and } \ \theta = \alpha^{-1}(\alpha - 2^{-1}).$$

This completes the proof in the complex case. Note that if $\alpha \to 1/2$ then $\theta \to 0$ and the known upper bound for $\phi(\theta)$ (see Theorem 7.11) yields that $C(\alpha)$ is $0((\alpha - 2^{-1})^{-1/2})$.

Let us know briefly check that the real case follows from the complex case. If E is an n-dimensional normed space over $\mathbf{R}$, we may consider a complexification $\widetilde{E}$ which is an n-dimensional normed space over $\mathbf{C}$ (hence $2n$-dimensional over $\mathbf{R}$) such that E embeds isometrically as a real subspace of $\widetilde{E}$ and such that there is an $\mathbf{R}$-linear contractive projection $P : \widetilde{E} \to E$. To obtain $\widetilde{E}$ we may for instance consider the space of all $\mathbf{R}$-linear operators from $\mathbf{C}$ into E (equivalently $\mathbf{C} \otimes E$) equipped with the operator norm.

By the complex case of Theorem 7.13, there is a **C**-linear isomorphism $u : \ell_2^n(\mathbf{C}) \to \widetilde{E}$ satisfying (7.36). Let $H = u^{-1}(E)$ (we view E as a subset of $\widetilde{E}$). Then $\dim_{\mathbf{R}} H = n$ and H may be identified with $\ell_2^n(\mathbf{R})$. Let $\widetilde{u} = Pu_{|H} : H \to E$. Note that $\widetilde{u}$ is an **R**-linear isomorphism. We have clearly $d_{2k-1}(\widetilde{u}) \leq d_k(u)$ and $c_{2k-1}(\widetilde{u}^{-1}) \leq c_k(u^{-1})$, where these numbers should be interpreted in the complex sense for u and in the real sense for $\widetilde{u}$. Therefore, if u satisfies (7.36) then $\widetilde{u}$ satisfies (7.36) (in the real sense) with C replaced by $2^\alpha C$ (the factor 2^α appears here to compensate the change from $2k - 1$ to k). ■

Second Proof of Theorem 7.1: Let B be a ball in $\mathbf{R}^n$ and let E be the associated normed space. Let u be as in Theorem 7.13. By Theorem 5.2 we have *a fortiori*

$$(7.40) \qquad \max\left\{e_n(u), e_n(u^{-1}), e_n(u^*), e_n(u^{-1*})\right\} \leq \lambda,$$

where we have set $\lambda = \rho_\alpha C(\alpha)$. Let $D = u(B_{\ell_2^n})$ be the ellipsoid associated to u. Equivalently, (7.40) may be rewritten as
$$(7.41)$$
$$\max\left\{N(B, \lambda D), N(D, \lambda B), N(B^\circ, \lambda D^\circ), N(D^\circ, \lambda B^\circ)\right\} \leq 2^{n-1}.$$

Hence *a fortiori* by (7.6) (with $A_1 = B$, $A_2 = \lambda D$)

$$(7.42) \qquad \mathrm{vol}\,(B + D)^{1/n} \leq 2(\lambda + 1)\,\mathrm{vol}\,(D)^{1/n}.$$

Similarly, we have

$$(7.43) \qquad \mathrm{vol}\,(B^\circ + D^\circ)^{1/n} \leq 2(\lambda + 1)\,\mathrm{vol}\,(D^\circ)^{1/n}.$$

On the other hand, by (7.41) and (7.7) we have

$$N(D, 2(\lambda B \cap D)) \leq 2^{n-1};$$

hence

$$(7.44) \qquad \mathrm{vol}\,(D)^{1/n} \leq 4(\lambda + 1)\,\mathrm{vol}\,(B \cap D)^{1/n}.$$

Similarly, we have

$$(7.45) \qquad \mathrm{vol}\,(D^\circ)^{1/n} \leq 4(\lambda + 1)\,\mathrm{vol}\,(B^\circ \cap D^\circ)^{1/n}.$$

Recollecting (7.42), (7.43), (7.44), and (7.45), we conclude that

$$M(B, D) \leq 64(\lambda + 1)^4. \quad ■$$

By Theorem 5.2, the following is an immediate corollary of Theorem 7.13.

Corollary 7.15. *For each $\alpha > 1/2$ there is a constant $C_1 = C_1(\alpha)$ such that for any n and any n-dimensional normed space there is an isomorphism $u : \ell_2^n \to E$ such that*

$$\forall \, k \geq 1 \quad \max\left\{e_k(u), e_k(u^*), e_k(u^{-1}), e_k(u^{-1*})\right\} \leq C_1(n/k)^{\alpha}.$$

Moreover, $C_1(\alpha)$ is $0((\alpha - 2^{-1})^{-1/2})$ when $\alpha \to 1/2$.

We can phrase this corollary in a more geometric manner as follows:

Corollary 7.16. *For every $p < 2$ there is a constant K_p such that for any n and any ball B in $\mathbf{R}^n$ there is an ellipsoid D in $\mathbf{R}^n$ such that for all $t \geq 1$ we have*

$$\max\left\{N(B, tD), N(D, tB), N(B^\circ, tD^\circ), N(D^\circ, tB^\circ)\right\} \leq \exp(K_p n t^{-p}).$$

Proof: This follows immediately from Corollary 7.15 (at least for $t \geq C_1$) applied to the normed space E associated to B, letting $D = u(B_{\ell_2^n})$ and $p = 1/\alpha$. Using Lemma 4.16 we see that by suitably adjusting the constant K_p we can obtain a similar estimate for all $t \geq 1$. ■

Remark: Let us briefly review here the improvements that Theorem 7.13 brings to corollaries 7.3 and 7.10. Fix a number $\alpha > 1/2$. Let B be a ball in $\mathbf{R}^n$ with associated normed space E. We will say that B is α-*regular* if we can choose the isomorphism u in Theorem 7.13 equal to a multiple of the identity on $\mathbf{R}^n$. Equivalently, the ellipsoid associated to u is a multiple of the canonical Euclidean ball. With this terminology, Theorem 7.13 says that *every ball is equivalent to a ball $\widetilde{B}$ which is α-regular*. (To verify this, we simply apply Theorem 7.13 to E and let $\widetilde{B} = u^{-1}(B)$.) Again, we emphasize that if B is α-regular, then its polar B° and all the multiples of B and B° are also α-regular. We may refine corollary 7.3 as follows: Let $B_1, \ldots, B_m$ be a finite set of α-*regular* balls in $\mathbf{R}^n$. Then we have
(7.46)
$$\mathrm{vol}\,(B_1 + \ldots + B_m)^{1/n} \leq \chi(\alpha) m^{\alpha} \left[\mathrm{vol}\,(B_1)^{1/n} + \ldots + \mathrm{vol}\,(B_m)^{1/n}\right],$$

where $\chi(\alpha)$ is a constant depending only on α. Indeed, let $D_1, \ldots, D_m$ be the ellipsoids associated to $B_1, \ldots, B_m$ according to Theorem 7.13.

By the proof of corollary 7.16 if $p = 1/\alpha$ there is a constant $K = K_p$ such that

(7.47) $\forall\, t \geq 1$ $\max\{N(B_i, t\, D_i), N(D_i, t\, B_i)\} \leq \exp(Knt^{-p}).$

Hence

$$N\,(B_1 + \ldots + B_m, t(D_1 + \ldots + D_m)) \leq \Pi\ N(B_i, t\, D_i)$$
$$\leq \exp(Knm\, t^{-p})$$

so that

$$\mathrm{vol}\,(B_1 + \ldots + B_m)^{1/n} \leq t \exp(Kmt^{-p})\ \mathrm{vol}\,(D_1 + \ldots + D_m)^{1/n}.$$

Now since we assume $B_1, \ldots, B_m$ α-regular, the ellipsoids $D_1, \ldots, D_m$ are all multiples of a fixed ellipsoid so that (trivially)

$$\mathrm{vol}\,(D_1 + \ldots + D_m)^{1/n} = \mathrm{vol}\,(D_1)^{1/n} + \ldots + \mathrm{vol}\,(D_m)^{1/n}.$$

Moreover, by (7.47) with $t = 1$ we have

$$\mathrm{vol}\,(D_i)^{1/n} \leq \exp(K)\ \mathrm{vol}\,(B_i)^{1/n},$$

hence we conclude that for all $t \geq 1$

$$\mathrm{vol}\,(B_1 + \ldots + B_m)^{1/n}$$
$$\leq t \exp(Kmt^{-p} + K)\left[\mathrm{vol}\,(B_1)^{1/n} + \ldots + \mathrm{vol}\,(B_m)^{1/n}\right].$$

Finally, choosing $t = m^{\alpha} = m^{1/p}$ we obtain (7.46). ∎

Concerning Corollary 7.10, if we assume $B_1, \ldots, B_m$ α-regular, we find

$$\mathrm{vol}\,(B_1 \cap \ldots \cap B_m)^{1/n}$$
$$\geq \delta(\alpha)m^{-1-\alpha}\inf\left\{\mathrm{vol}\,(B_1)^{1/n}, \ldots, \mathrm{vol}\,(B_m)^{1/n}\right\},$$

for some positive constant $\delta(\alpha)$ depending only on $\alpha > 1/2$. Indeed, this is easy to deduce from (7.46) and Corollary 7.2.

Remarks:

(i) In the proof of Theorem 7.13, we can use Theorem 5.14 instead of Theorem 5.8; we then obtain the same result but only for all

$\alpha > 1$. Of course, this suffices to obtain (7.40) and hence to complete the second proof of Theorem 7.1.

(ii) By a simple modification of the proof of Theorem 7.13 one can give a direct proof of Corollary 7.15 using the entropy numbers $e_k(u)$ instead of $d_k(u)$ and $d_k(u^{-1})$. The ingredients for this alternate route are reduced to the fact that

$$\sup_{k \geq 1} k^{1/2} \max \{e_k(u), e_k(u^*)\} \leq K \, \ell(u)$$

(for some absolute constant K) and the fact that the entropy numbers satisfy both inequalities (7.27) and (7.28).

Remark: It is worthwhile to mention that Theorem 7.13 is invalid for $\alpha < 1/2$. Indeed, by Theorem 5.5 and Theorem 5.2 we have for all $u : \ell_2^n \to E$

$$\forall \, \alpha < 1/2 \quad \ell(u) \leq \psi(\alpha) n^{1/2} \sup_{k \geq 1}(k/n)^{\alpha} c_k(u^*),$$

for some constant $\psi(\alpha)$ depending only on α. Therefore, if $\alpha < 1/2$ then (7.36) implies

$$\ell(u) \leq \psi(\alpha) C n^{1/2} \text{ and } \ell(u^{-1*}) \leq \psi(\alpha) C n^{1/2}.$$

But it is well known that this is impossible, for example, if $E = \ell_\infty^n$ or ℓ_1^n; actually, it is not difficult to check that there is a number $\delta > 0$ such that for all n and all isomorphisms $u : \ell_2^n \to \ell_\infty^n$ we have

(7.48) $$\ell(u)\ell(u^{-1*}) \geq \delta n (\text{Log } n)^{1/2}.$$

This follows from the following facts: For some constant K_1 we have

(i) $$\sup_{k \geq 1}(\text{Log } k)^{1/2} a_k(u) \leq K_1 \ell(u).$$

For some constants K_2 and K_3 we have, for all $v : \ell_\infty^n \to \ell_2^n$,

(ii) $$\sup_k k^{1/2} a_k(v) \leq K_2 \ell(v^*)$$

and

(iii) $$\pi_2(v) \leq K_3 \|v\|.$$

The first point (i) follows easily from Lemma 1.8, (4.14) and (4.18). The second point (ii) follows from the fact that ℓ_1 is of cotype 2 (see the subsequent definition 10.1), while the third point (iii) follows from a weak form of Grothendieck's theorem (cf. e.g. Theorem 4.3 in [P2], p. 54).

Then we deduce from (iii) that for all subspaces $F \subset \ell_2^n$ with $\dim F = d$ we have

$$d^{1/2} = \pi_2(I_F) \leq \|u_{|F}\|\pi_2\left(P_F u^{-1}\right)$$
$$\leq \|u_{|F}\|\|K_3\|P_F u^{-1}\|.$$

This implies immediately

$$(n/2)^{1/2} \leq a_{n/4}(u)a_{n/4}(u^{-1})K_3;$$

hence, by (i) and (ii) above,

$$(n/2)^{1/2}\left(\text{Log } n/4\right)^{1/2}(n/4)^{1/2} \leq K_1 K_2 K_3\, \ell(u)\ell\left(u^{-1*}\right),.$$

which proves (7.48). We thus conclude that Theorem 7.13 fails for $\alpha < 1/2$. In the case $\alpha = 1/2$, we do not know, but we suspect that at least some additional logarithmic term should be necessary in (7.36) in that case also.

Notes and Remarks

The main result—Theorem 7.1—is due to Milman [M5], but our proofs are different and simpler than the original one. (Actually, the details of Milman's original proof have never been published.) In our opinion, this major result is the culminating point of an investigation which was carried on in several steps during the period 1984—85. Chronologically, the main steps were [M1] then [BM], and finally [M5]. (Our exposition does not match this chronology.)

One advantage of our proof is that it does not assume knowledge of any of the papers [BM] or [M1], so that we can recover their results as immediate corollaries. However, the basic ingredients in our proof are quite similar to those which are used in all these papers. Corollary 7.2 is due to Bourgain and Milman [BM]. (Of course, it was actually proved before Theorem 7.1, but after Corollary 7.9.) Let us explain a little bit the genesis of this result.

Let B be any ball in $\mathbf{R}^n$. Let $s(B) = (\text{vol}(B)\,\text{vol}(B^\circ))^{1/n}$. Around 1980/81, Saint-Raymond [StR1] found a new proof of the Santaló inequality

$$s(B) \leq s(B_{\ell_2^n})$$

and also of the equality case (equality holds iff B is an ellipsoid).

This suggested studying the converse inequality. In 1939, Mahler conjectured that the following lower bound always holds:

$$(7.49) \qquad s(B) \geq s(B_{\ell_\infty^n}) = (4^n (n!)^{-1})^{1/n}.$$

This was established for the balls of spaces with a 1-unconditional basis (cf. [StR1]) and for zonoids (cf. [Re1], [Re3] and [GMR]). Moreover, for spaces with a 1-unconditional basis Mathieu Meyer proved in [Me] that equality can occur in (7.49) only if the normed space E associated to B is a *mixture* of ℓ_∞^k's and ℓ_1^m's, we refer to [Me] for more precision. (A different proof of this was given by Reisner [Re2].)

In full generality (7.49) is still an open problem; however, a simple computation shows that there is a numerical constant $\alpha > 0$ such that $s(B_{\ell_\infty^n}) \geq \alpha s(B_{\ell_2^n})$, so that the conjecture (7.49) implies

$$(7.50) \qquad s(B) \geq \alpha s(B_{\ell_2^n})$$

with $\alpha > 0$ independent of n.

Motivated by Saint-Raymond's work, Gordon and Reisner (and, independently, the author) observed that if E is the normed space associated to B, we have

$$(7.51) \qquad s(B) \geq K(E)^{-1} s(B_{\ell_2^n})$$

and therefore (by Theorem 2.5)

$$(7.52) \qquad s(B) \geq K^{-1}(1 + \text{Log } d(E, \ell_2^n))^{-1} s(B_{\ell_2^n}).$$

In particular, this immediately yields (for some numerical constant $\alpha' > 0$)

$$(7.53) \qquad s(B) \geq \alpha'(1 + \text{Log } n)^{-1} s(B_{\ell_2^n})$$

which was already a significant improvement over previous estimates (apparently only $n^{-1/2}$ in place of $\alpha'(1 + \text{Log } n)^{-1}$ was known).

We note in passing that (7.51) is very easy to derive from the properties of the ℓ-ellipsoid. Indeed, consider $u : \ell_2^n \to E$ with $\ell(u) = n^{1/2}$ and $\ell(u^{-1*}) \leq K(E)n^{1/2}$. Let $D = u(B_{\ell_2^n})$. As seen in Chapter 6, we have

$$\left(\frac{\text{vol}(B)}{\text{vol}(D)}\right)^{1/n} = \left(\int \|ux\|^{-n}\, d\sigma(x)\right)^{1/n};$$

hence, by convexity,

$$\geq \left(\int \|ux\|\, d\sigma(x)\right)^{-1} \geq \ell(u)^{-1} n^{1/2}.$$

Similarly,

$$\left(\frac{\text{vol}(B^\circ)}{\text{vol}(D^\circ)}\right)^{1/n} \geq \ell(u^{-1*}) n^{1/2}.$$

Therefore, $s(B) \geq s(D)n(\ell(u)\ell(u^{-1*}))^{-1} = s(D)K(E)^{-1}$, and this proves (7.51), since $s(D)$ is independent of the choice of the ellipsoid D.

We now come to the work of Bourgain and Milman. They proved (7.50) using (7.52) and a clever iteration argument in order to get rid of the Log n factor in (7.53).

This was part of an ambitious program proposed by Milman following his proof of the QS-Theorem in [M1]. Indeed, as we will see in the next chapter, the situation for the QS-Theorem was analogous; it was first proved with a Log n factor in [M3] (cf. Theorem 8.2) and later in [M1] this Log n factor was removed by an iteration argument (cf. Theorem 8.4 and its proof). After that, Milman proposed to systematically review all the situations where an apparently unnecessary Log n term appeared in a similar way and to try an iteration procedure. There is still research in progress along these lines. See [M7] for a survey.

Corollary 7.3 comes from [M5].

Lemma 7.6 is a simple variation on a lemma appearing in [MP1] and Proposition 7.7 is a simple consequence of it, modulo known results. Corollary 7.8 is due to König and Milman [KM]. Corollary 7.9 is the QS-Theorem which first appeared in [M1]. We give a direct proof (following [M1]) in the next chapter. Finally, Corollary 7.10 is an obvious dualization of Corollary 7.3.

More recently, Milman (cf. [M8] and [M9]) found another proof of his Theorem 7.1, partially influenced by our first proof included above; his proof is based roughly on similar ideas, but nevertheless is different.

Even more recently (February 1988), I found a new approach (cf. [P8]) based on interpolation theory. This is included above as the "second proof" of Theorem 7.1. This approach actually yields several improvements over Theorem 7.1, stated as Theorem 7.13 and Corollaries 7.15 and 7.16, which all come from [P8]. In this approach, the main new ingredient is Theorem 7.11, which was first proved in [P9]. (Somehow, this was a preliminary step on the way to the later results of [P3].) Lemma 7.12 is an elementary known fact from the Theory of Functions of one complex variable.

Chapter 8

Another Proof of the QS-Theorem: The Iteration Procedure

This chapter is independent of the previous one. We show here a proof of the QS-Theorem directly deduced from Theorem 5.8 and an iteration argument. We then deduce the inverse Santaló inequality [BM] directly from the QS-Theorem.

Let us first reformulate Theorem 5.8.

Theorem 8.1. *Let* $u : \ell_2^n \to E$ *be an operator. For all subspaces* $H \subset \ell_2^n$ *we denote by* $u_H : H \to u(H)$ *the restriction of* u *(we consider* $u(H)$ *as a normed subspace of* E*). For any* $k \geq 1$*, let* $\widetilde{d}_k(u) = \sup_{H \subset \ell_2^n} d_k(u_H)$. *Then we have*

$$\sup k^{1/2} \widetilde{d}_k(u) \leq C_1 \ell(u)$$

for some numerical constant C_1.

Proof: This is a trivial consequence of Theorem 5.8, once one observes that $\ell(u_H) \leq \ell(u)$. ∎

We now combine this result with the results of Chapter 3 to prove a weak form of the QS-Theorem. We will denote by $QS(E)$ the class of all quotient spaces of a subspace of E. In other words, the elements of $QS(E)$ are all the normed spaces of the form E_1/E_2 for some subspaces $E_2 \subset E_1 \subset E$. The reader should note that the class $QS(E)$ coincides with the class of all subspaces of a quotient of E, which we might denote by $SQ(E)$; thus the identity $QS(E) = SQ(E)$ is elementary. In particular, if $F \in QS(E)$, then all subspaces and all quotient spaces of F belong to $QS(E)$, and we have $QS(F) \subset QS(E)$. Therefore, there is no need to consider more general classes such as *subspaces of quotient spaces of subspaces of* E since these are identical to $QS(E)$.

We recall the notation $d_F = d(F, \ell_2^d)$ if $\dim F = d$.

127

Theorem 8.2.   *There is a constant C such that, for all n and for all normed spaces E with $\dim E = n$, for every integer k with $1 \le k \le n$, there is an F in $QS(E)$ such that $\dim F > n - k$ and $d_F \le C(n/k)K(E)$.*

Proof: By Theorem 3.9, there is an isomorphism $u : \ell_2^n \to E$ such that

(8.1) $\ell(u) = n^{1/2}$ and $\ell((u^{-1})^*) \le n^{1/2}K(E)$.

By Theorem 5.8, there is a numerical constant C_1 such that

$$k^{1/2}c_k(u^{-1}) \le C_1\ell((u^{-1})^*).$$

Hence $\exists\, E_1 \subset E$ with $\operatorname{codim}E_1 < k$ such that

(8.2) $\|u_{|E_1}^{-1}\| \le C_1 k^{-1/2}\ell((u^{-1})^*).$

Let $H = u^{-1}(E_1) \subset \ell_2^n$. Consider u_H as in Theorem 8.1; we have

$$d_k(u_H) \le \tilde{d}_k(u) \le C_1 k^{-1/2}\ell(u).$$

Hence, $\exists\, E_2 \subset E_1$ with $\dim E_2 < k$ such that, if $Q : E_1 \to E_1/E_2$ denotes the quotient map, we have

$$\|Qu_H\| \le C_1 k^{-1/2}\ell(u) = C_1(n/k)^{1/2}.$$

We claim that E_1/E_2 (which is clearly in $QS(E)$) satisfies

$$d_{E_1/E_2} \le C_1^2(n/k)K(E).$$

Indeed, consider the operator $Qu_H : H \to E_1/E_2$. Let $\tilde{H} = (\operatorname{Ker}(Qu_H))^{\perp}$. Then Qu_H factors canonically as $H \xrightarrow{P} \tilde{H} \xrightarrow{v} E_1/E_2$ where P is the orthogonal projection onto $\tilde{H}$ and where v is the restriction of Qu_H to $\tilde{H}$. Clearly, v is an isomorphism and $\|v\| \le \|Qu_H\| \le C_1(n/k)^{1/2}$.

$$E$$

$$\cup$$

$$
\begin{array}{ccccc}
H & \xrightarrow{u} & E_1 & \xrightarrow{u^{-1}} & H \\
 & & \downarrow{Q} & & \downarrow{P} \\
 & & E_1/E_2 & \xrightarrow{v^{-1}} & \tilde{H}
\end{array}
$$

On the other hand (see the above diagram), $u^{-1}|_{E_1} : E_1 \to H$ has norm majorized by (8.2), hence the same is true for $Pu^{-1}|_{E_1} : E_1 \to \widetilde{H}$ which vanishes on E_2. Passing to the quotient by E_2, we find an operator from E_1/E_2 onto $\widetilde{H}$ which coincides with v^{-1}. Hence, we conclude by (8.1) and (8.2)

$$\|v^{-1}\| \le \|u^{-1}|_{E_1}\| \le C_1 \, k^{-1/2} K(E) n^{1/2},$$

so that finally

$$d_{E_1/E_2} \le \|v\| \, \|v^{-1}\| \le C_1^2 (n/k) K(E).$$

Since $\dim E_1/E_2 > n - 2k + 1$, we thus obtain Theorem 8.2 for odd integers; the case of even integers follows immediately by adjusting the constant C. ∎

Remark 8.3: As an immediate consequence of Theorem 8.2 and 3.10, we obtain the following weak form of the QS-Theorem: there is a constant C such that for all n and all n-dimensional spaces E, for each $0 < \delta < 1$, there is a space $F \in QS(E)$ with $\dim F \ge \delta n$ such that $d_F \le C(1 - \delta)^{-1}(1 + \text{Log } d_E)$.

The factor $(1 + \text{Log } d_E)$ is the weak point of this estimate. It is rather surprising that it can be *automatically* replaced by a factor of the form $1 + \text{Log}(C(1-\delta)^{-1})$ which is much better (it does not depend on E). This can be done by a trick discovered by Milman (we call it the iteration argument). It can be stated as an elementary real-variable lemma (Lemma 8.6 below). This lemma will immediately yield the following improvement of Remark 8.3.

Theorem 8.4. *There is a numerical constant C' such that for all n and all n-dimensional normed spaces E, we have $\forall\, k = 1, 2, \dots, n$, $\exists\, F \in QS(E)$ with $\dim\, F = n - k$ such that*

$$d_F \le C'(n/k) \text{Log}(C'n/k).$$

Remark: In other words, for each $k = 1, \dots, n$ there are subspaces $F_2 \subset E_1 \subset E$ with $\dim E_1 - \dim F_2 = n - k$ such that

$$d_{E_1/F_2} \le C'(n/k) \text{Log}(C'n/k).$$

It is easy to see that we may identify E_1/F_2 with the vector space $E_1 \cap F_2^\perp = E_1 \ominus F_2$ equipped with the unit ball $P_{F_2^\perp}(B_{E_1}) = P_{F_2^\perp}(B \cap E_1)$

(cf. Chapter 1). Letting $E_2 = E_1 \ominus F_2$, we finally identify E_1/F_2 with E_2 equipped with the ball $P_{E_2}(B \cap E_1)$. Then, we can reformulate geometrically the previous statement as follows: Let $\lambda = C'(n/k)\operatorname{Log}(C'n/k)$; there is an $n - k$-dimensional ellipsoid $D \subset E_2$ such that

$$D \subset P_{E_2}(B \cap E_1) \subset \lambda D.$$

We come now to the *iteration lemmas* which allow us to pass from Remark 8.3 to Theorem 8.4.

Lemma 8.5. *Let $f : [0, 1[\to \mathbb{R}_+$ be a bounded function satisfying for some $K \geq 0, C \geq 1$ and $0 < \theta < 1$;*

$$\forall \, \delta \in [0, 1[\qquad f(\delta^2) \leq C(1 - \delta)^{-K} f(\delta)^{\theta}.$$

Then we have

$$\forall \, \delta \in [0, 1[\qquad f(\delta) \leq C_{\theta}(1 - \delta)^{-\frac{K}{1-\theta}},$$

where $C_{\theta} = C^{(1-\theta)^{-1}} 2^{K(1-\theta)^{-2}}$.

Lemma 8.6. *Let $f : [0, 1[\to \mathbb{R}_+$ be a bounded function such that $f(t) \geq 1$ for all t and satisfying for some constant $K' \geq 1$*

$$(8.3) \qquad \forall \, \delta \in [0, 1[\quad f(\delta^2) \leq K'(1 - \delta)^{-1} \operatorname{Log}(e^2 f(\delta)).$$

Then

$$\forall \, \delta \in [0, 1[\quad f(\delta) \leq \frac{2^5 e K'}{1 - \delta} \operatorname{Log}|e K'(1 - \delta)^{-1}|.$$

Proof of Lemma 8.5:

   1. Particular case: $C = 1, K = 0$. Then for all δ in $[0, 1[$ such that $\delta = x^{2^n}$ for some x in $[0, 1[$, we have $f(x^2) \leq f(x)^{\theta}, f(x^4) \leq f(x^2)^{\theta} \leq f(x)^{\theta^2}$, etc., so that $f(\delta) \leq f(x)^{\theta^n}$.

   Since f is assumed bounded by some number M, we have $f(\delta) \leq M^{\theta^n}$. Since n can be taken arbitrarily large, we find (since $M^{\theta^n} \to 1$ when $n \to \infty$) that $f(\delta) \leq 1$ for all δ in $[0, 1[$ as announced.

   2. General case: We reduce it to the particular case by defining φ as

$$\varphi(\delta) = (C_{\theta})^{-1}[(1 - \delta)^{K(1-\theta)^{-1}}] f(\delta).$$

Note that $1 - \delta^2 \leq 2(1 - \delta)$, so that one easily checks

$$\varphi(\delta^2) \leq (C_{\theta})^{-1}[(2(1 - \delta))^{K(1-\theta)^{-1}}] f(\delta^2),$$

and the value of C_θ is adjusted so that we obtain

$$\varphi(\delta^2) \le \varphi(\delta)^\theta.$$

Clearly, φ is also bounded, hence by the first part of the proof $\varphi(\delta) \le 1$ for all $0 \le \delta < 1$. ∎

Proof of Lemma 8.6: We use the elementary formula

$$(8.4) \qquad \forall y \ge e^2 \quad \inf_{0<\theta<\frac{1}{2}} \frac{1}{\theta}y^\theta = e \text{ Log } y$$

(the infimum is attained for $\theta = (\text{Log } y)^{-1} < \frac{1}{2}$).

The assumption of Lemma 8.6 can thus be equivalently formulated as

$$\forall \delta \in [0,1[\quad f(\delta^2) \le K'(1-\delta)^{-1}(e\theta)^{-1}(e^2 f(\delta))^\theta \text{ for all } 0 < \theta < \frac{1}{2}.$$

By Lemma 8.5, we obtain

$$(8.5) \qquad f(\delta) \le A_\theta(1-\delta)^{-(1-\theta)^{-1}},$$

with

$$A_\theta = \left\{ 2^{\frac{1}{1-\theta}} \cdot \frac{K'e^{2\theta}}{e\theta} \right\}^{\frac{1}{1-\theta}}.$$

Using the inequality $\frac{1}{1-\theta} \le 1 + 2\theta$ valid if $0 < \theta < \frac{1}{2}$, we obtain

$$A_\theta \le B_\theta \text{ with } B_\theta = 2^4 \left(\frac{K'e^{2\theta}}{e\theta} \right) \left(\frac{K'}{\theta} \right)^{2\theta}.$$

Note that

$$\left(\frac{1}{\theta} \right)^{2/e} \le e^{2/e} \le e \text{ if } 0 < \theta < \frac{1}{2},$$

hence $B_\theta \le 2^4 \frac{K'}{\theta}(K'e)^{2\theta}$ so that (8.5) implies

$$f(\delta) \le \frac{2^4 K'}{1-\delta} \frac{1}{\theta} \left[\frac{K'^2 e^2}{(1-\delta)^2} \right]^\theta.$$

Using (8.4) once again, we obtain

$$f(\delta) \le 2^4 K'(1-\delta)^{-1}e \text{ Log } \left(\frac{K'^2 e^2}{(1-\delta)^2} \right). \quad ∎$$

Proof of Theorem 8.4: Let E be an n-dimensional space. Let $f(\delta)$ be defined as

$$f(\delta) = \inf\{d_F, F \in QS(E), \dim F \geq \delta n\}.$$

We claim that (by Remark 8.3) this function satisfies the hypothesis of Lemma 8.6 for some numerical constant K'. Indeed, given F_1 in $QS(E)$ with $\dim F_1 \geq \delta n$, we can apply Remark 8.3 to the space F_1, obtaining F_2 in $QS(F_1)$ with $\dim F_2 \geq \delta \dim F_1$ such that $d_{F_2} \leq C(1 - \delta)^{-1}(1 + \mathrm{Log}\, d_{F_1})$. Taking the infimum over F_1, we find $f(\delta^2) \leq C(1 - \delta)^{-1}(1 + \mathrm{Log}\, f(\delta))$ and we can find a constant K' such that (8.3) holds.

By Lemma 8.6, we then obtain the statement of Theorem 8.4 (take $\delta = 1 - \frac{k}{n}$). ∎

In the second part of this chapter, we show that the inverse Santaló inequality (first proved in [BM]) is a direct consequence of the QS-Theorem.

Theorem 8.7. *There are numerical constants $\alpha > 0$ and $\beta > 0$ such that, for all n, any ball $B \subset \mathbf{R}^n$ satisfies*

$$\alpha \leq \left(\frac{\mathrm{vol}(B)\,\mathrm{vol}(B^o)}{\mathrm{vol}(B_{\ell_2^n})^2} \right)^{1/n} \leq \beta.$$

Remark: We recall that Santaló [Sa1] proved the upper bound with $\beta = 1$, but this does not seem to be possible by our methods. The proof is based on the following known result:

Lemma 8.8. *Let B be a ball in $\mathbf{R}^n$, let $S \subset \mathbf{R}^n$ be a k-dimensional subspace. We denote by P_S the orthogonal projection from $\mathbb{R}^n$ onto S. We have*
(8.6)
$$\binom{n}{k}^{-1} \mathrm{vol}(S \cap B)\,\mathrm{vol}(P_{S^{\perp}}(B)) \leq \mathrm{vol}(B) \leq \mathrm{vol}(S \cap B)\,\mathrm{vol}(P_{S^{\perp}}(B)).$$

Recall here that
$$\binom{n}{k} = \frac{n!}{k!(n-k)!} \leq 2^n.$$

Proof: We clearly have (denoting by dx Lebesgue measure on $S^{\perp}$)

$$\mathrm{vol}(B) = \int_{S^{\perp}} \mathrm{vol}((x + S) \cap B)\, dx$$

(here $\text{vol}((x + S) \cap B)$ is the k-dimensional volume). Since for any x in $S^{\perp}, (x + S) \cap B \neq \phi$ implies $x \in P_{S^{\perp}}(B)$, we have

$$(8.7) \qquad \text{vol}(B) = \int_{P_{S^{\perp}}(B)} \text{vol}((x + S) \cap B)\, dx.$$

By the Brunn–Minkowski inequality, we have

$$(8.8) \qquad \text{vol}((x + S) \cap B) \leq \text{vol}(S \cap B).$$

(Indeed, let $A_1 = (x + S) \cap B$, $A_2 = (-x + S) \cap B$, then

$$\text{vol}(A_1)^{1/k} = \frac{1}{2}((\text{vol}(A_1)^{1/k} + (\text{vol}(A_2))^{1/k}) \leq \text{vol}(2^{-1}(A_1 + A_2))^{1/k}$$

$$\leq \text{vol}(S \cap B)^{1/k}.)$$

Therefore, the right side of (8.5) follows immediately from (8.7) and (8.8).

Conversely, let $t(x)$ be the gauge of $P_{S^{\perp}(B)}$. We will show that if $x \in P_{S^{\perp}(B)}$ then:

$$(8.9) \qquad \text{vol}((x + S) \cap B) \geq (1 - t(x))^k \text{vol}(S \cap B).$$

Indeed, assume $x \in tP_{S^{\perp}}(B)$ with $0 \leq t \leq 1$, then $\exists\, b \in B$ such that $x - tb \in \mathbf{S}$. It follows that

$$(S + x) \cap B \supset (1 - t)(S \cap B) + t\, b;$$

hence, $\text{vol}((S + x) \cap B) \geq (1 - t)^k \text{vol}(S \cap B)$, which establishes (8.9). Now we conclude from (8.7) that

$$(8.10) \qquad \text{vol}(B) \geq \int_{P_S^{\perp}(B)} (1 - t(x))^k\, dx\ \text{vol}(S \cap B).$$

Let $K = P_{S^{\perp}}(B)$; it is easy to check that

$$\int_K (1 - t(x))^k\, dx = \int_0^1 (1 - t)^k\, d(\text{vol}(tK))$$

$$= \int_0^1 (1 - t)^k\, dt^{n-k}\, \text{vol}(K) = \binom{n}{k}^{-1} \text{vol}(K).$$

Therefore, (8.10) implies the left side of (8.6). ■

Remark: At the cost of an extra factor 2^n, we can prove the right side of (8.5) without invoking the Brunn–Minkowski inequality. This makes the proof of (8.5) (and hence of Theorem 8.7) even more elementary.

Proof of Theorem 8.7: Let us denote $s(B) = (\text{vol}(B)\,\text{vol}(B^o))^{1/n}$. We recall that this is an *affine* invariant, i.e. for any $u : \mathbf{R}^n \to \mathbf{R}^n$ linear and invertible we have $s(B) = s(u(B))$. Let N be an integer; we denote

$$s_N = \inf\{s(B)\left(s(B_{\ell_2^n})\right)^{-1}, n \leq N, B \subset \mathbf{R}^n\}$$
$$S_N = \sup\{s(B)\left(s(B_{\ell_2^n})\right)^{-1}, n \leq N, B \subset \mathbf{R}^n\}.$$

We will show that $\inf_N s_N > 0$ and $\sup_N S_N < \infty$. We first give the proof for s_N. Consider $n \leq N$ and a ball $B \subset \mathbf{R}^n$. Let $k = \left[\frac{n}{2}\right]$. By the remark following Theorem 8.4, there are subspaces $E_2 \subset E_1 \subset \mathbf{R}^n$ with $\dim E_2 = n - k \geq \frac{n}{2}$ and an ellipsoid D included in E_2 such that

$$D \subset P_{E_2}(E_1 \cap B) \subset K\,D,$$

where $K \geq 1$ is a numerical constant.

This clearly implies

(8.11) $$K^{-1}s(D) \leq s(P_{E_2}(E_1 \cap B)) \leq K\,s(D),$$

and since $s(B)$ is an affine invariant, we have

$$s(D) = s(B_{\ell_2^{n-k}}).$$

Let us denote $V_k = \text{vol}(B_{\ell_2^k})$, so that $s(B_{\ell_2^k}) = (V_k)^{2/k}$. Note that (8.6) implies in particular

(8.12) $$V_n \leq V_k V_{n-k}.$$

Applying now (8.6) to B, we obtain

$$\text{vol}(B) \geq 2^{-n}\,\text{vol}(E_1 \cap B)\,\text{vol}(P_{E_1^\perp}(B)),$$

and applying it one more time, we find

$$\text{vol}(B) \geq 2^{-2n}\,\text{vol}(P_{E_2}(E_1 \cap B))\,\text{vol}(E_2^\perp \cap E_1 \cap B)\,\text{vol}(P_{E_1^\perp}(B)).$$

Proceeding similarly for B^o, we find

$$\text{vol}(B^o) \geq 2^{-2n} \text{vol}(E_2 \cap P_{E_1}(B^o)) \text{vol}(P_{E_2^{\perp}} P_{E_1}(B^o)) \text{vol}(E_1^{\perp} \cap B^o).$$

Let

$$A_1 = P_{E_1^{\perp}}(B) \quad \text{and} \quad A_2 = E_2^{\perp} \cap E_1 \cap B,$$

and $d_1 = \dim E_1^{\perp}$, $d_2 = \dim E_1 \cap E_2^{\perp}$. Note that $k = d_1 + d_2$. Taking the product of the last two inequalities, we obtain

$$s(B)^n \geq 2^{-4n}(s(P_{E_2}(E_1 \cap B)))^{n-k} s(A_1)^{d_1} s(A_2)^{d_2};$$

hence, by (8.11) and the definition of s_N,

$$s(B)^n \geq 2^{-4n}(K^{-1})^{n-k}(V_{n-k}^2)(s_N)^{d_1} V_{d_1}^2 (s_N)^{d_2} V_{d_2}^2;$$

therefore by (8.12), since $k = d_1 + d_2$,

$$\frac{s(B)}{(V_n^2)^{1/n}} \geq 2^{-4}(K^{-1})^{1-\frac{k}{n}}(s_N)^{\frac{k}{n}};$$

and since $s_N \leq s_1 = 1$ and $k \leq n/2$, we have $(s_N)^{k/n} \geq (s_N)^{1/2}$, so that we have proved

$$s_N = \inf\{s(B)(V_n^2)^{-1/n}\} \geq 2^{-4} K^{-1}(s_N)^{1/2}.$$

Dividing by $(s_N)^{1/2}$ and squaring we finally obtain

$$s_N \geq 2^{-8} K^{-2}.$$

(Note that we need to know that $s_N > 0$ for all N, but this is clear since by John's Theorem we have *a priori* $s_N \geq N^{-1/2}$.) This shows the lower bound for s_N; we obtain the upper bound by an entirely similar reasoning left as an exercise for the reader.

We note the following immediate consequence.

Corollary 8.9. *Let B_1, B_2 be balls in $\mathbf{R}^n$. Then (with $\alpha > o$ and $\beta > o$ as in Theorem 8.7) we have*

$$\alpha\beta^{-1} \leq \left(\frac{\text{vol}(B_1)\,\text{vol}(B_1^o)}{\text{vol}(B_2)\,\text{vol}(B_2^o)}\right)^{1/n} \leq \beta\alpha^{-1};$$

hence

$$\alpha\beta^{-1}\left(\frac{\text{vol}(B_2^o)}{\text{vol}(B_1^o)}\right)^{1/n} \leq \left(\frac{\text{vol}(B_1)}{\text{vol}(B_2)}\right)^{1/n} \leq \beta\alpha^{-1}\left(\frac{\text{vol}(B_2^o)}{\text{vol}(B_1^o)}\right)^{1/n}.$$

To conclude this chapter, we also indicate how a result of König–Milman [KM] can be deduced from Theorem 8.7.

Theorem 8.10. *There is a numerical constant c such that for all balls B_1, B_2 in $\mathbf{R}^n$ we have*

$$\frac{1}{c^n} N(B_2^o, B_1^o) \le N(B_1, B_2) \le c^n N(B_2^o, B_1^o).$$

Proof: It is easy to check (see (7.7) above) that

$$N(B_1, 2(B_1 \cap B_2)) \le N(B_1, B_2);$$

hence by (7.5),

$$\left(\frac{\mathrm{vol}(B_1)}{\mathrm{vol}(B_1 \cap B_2)}\right) \le 2^n N(B_1, B_2).$$

By Theorem 8.7,

$$\left(\frac{\mathrm{vol}((B_1 \cap B_2)^o)}{\mathrm{vol}(B_1^o)}\right) \le (2\beta\alpha^{-1})^n N(B_1, B_2);$$

hence (by Lemma 7.5),

$$N((B_1 \cap B_2)^0, B_1^o) \le (6\beta\alpha^{-1})^n N(B_1, B_2),$$

and since $B_2^o \subset (B_1 \cap B_2)^o$, we find

$$N(B_2^o, B_1^o) \le (6\beta\alpha^{-1})^n N(B_1, B_2).$$

The inverse inequality follows by exchanging the roles of (B_1, B_2) and (B_2^o, B_1^o). ■

This result immediately implies the following.

Corollary 8.11. *There is a numerical constant C such that for all n, all Banach spaces X, Y and all operators $u : X \to Y$ of rank n we have*

(8.13) $e_{[Cn]}(u^*) \le 2\, e_n(u).$

Proof: We may clearly assume (using Proposition 5.1) that $\dim X = \dim Y = n$. In that case the result follows (without the factor 2) from Theorem 8.10 with $B_1 = u(B_X)$ and $B_2 = B_Y$. The extra factor 2 in (8.13) comes from the use of Proposition 5.1.

Notes and Remarks

The main result of this chapter is the *QS-Theorem* of Milman. We follow the basic approach of [M1] except that we use at each step the improved estimates of [PT1], instead of the previous estimates of [M2] [M4].

Theorem 8.1 is a trivial reformulation of the main result of [PT1]. Theorem 8.2 first appeared in [M3], while Theorem 8.4 appeared in [M1], with a worse dependence on n/k. It was proved with $(n/k)^2$ instead of (n/k) in [M4]. The (n/k) estimate that we gave in Theorem 8.1 comes from [PT1].

Lemmas 8.5 and 8.6 are essentially in [M3] or [M2]. See also [D], [DS]. As already discussed in the Notes and Remarks following Chapter 7, Theorem 8.7 is due to Santaló [Sa1] for the upper bound (with $\beta = 1$) and to Bourgain and Milman [BM] for the lower bound.

The proof of Theorem 8.7 as a simple consequence of the QS-Theorem and Lemma 8.8 is new and is published here for the first time.

Lemma 8.8 is due to Rogers and Shephard [RSh] but the simple proof given above can be found in [Ch]. Finally, Theorem 8.10 and its Corollary are due to Hermann König and Milman [KM].

Chapter 9

Volume Numbers

This chapter is mainly based on [MP2].

Let K be a compact subset of a Hilbert space H. For $n \geq 1$, we define the *n-th volume number* of K as

(9.1)
$$v_n(K) = \sup \left\{ \left(\frac{\text{vol}(P(K))}{\text{vol}(P(B_H))} \right)^{1/n} \right\},$$

where the supremum runs over all orthogonal projections $P : H \to H$ with rank equal to n. If $\dim H < n$, we set $v_n(K) = 0$.

Let $T : X \to H$ be a compact operator defined on a Banach space X. We define the *volume numbers* of T as

$$v_n(T) = v_n(\overline{T(B_X)}).$$

In (9.1) above, vol denotes the volume in $P(H)$ (the range of P) equipped with the Euclidean structure induced by H. In particular, $\text{vol}(P(B_H))$ is equal to the volume of the Euclidean ball of $\mathbf{R}^n$ which we have denoted earlier by V_n. We have clearly $v_1(T) = \|T\|$. These numbers behave very much like the entropy numbers, as we shall see.

We first note that $\{v_n(K)\}$ is a non-increasing sequence. Indeed, this follows from the famous inequalities of Fenchel and Alexandrov on the mixed volumes (cf. [Al]). Let B be a compact subset of $\mathbf{R}^n$. We define

$$w_k(B) = \left(\int \frac{\text{vol}(P(B))}{V_k} \, dP \right)^{1/k},$$

where the integral is with respect to the uniform probability on the set of all the orthogonal projections $P : \mathbf{R}^n \to \mathbf{R}^n$ of rank k.

Clearly,

$$w_n(B) = \left(\frac{\text{vol}(B)}{V_n} \right)^{1/n}$$

and

$$w_1(B) = \int \sup_{t \in B} t(x) \, d\sigma(x),$$

where σ is the normalized area measure on the Euclidean sphere (i.e. the unit sphere of ℓ_2^n).

For a convex body B, $w_{n-1}(B)$ can be identified with the ratio

$$\left(\frac{a(B)}{a(B_{\ell_2^n})}\right)^{\frac{1}{n-1}}$$

of the *areas* of the surfaces respectively of B and of the Euclidean unit ball. The inequalities of Alexandrov simply say that the sequence $\{w_k(B)\}$ is a non-increasing sequence. We refer to [Al], [BF], [BZ], or [Eg] for a proof.

In particular, it is worthwhile to observe that this implies

$$w_n(B) \leq w_1(B) \quad \text{(Urysohn's inequality)}$$

and

$$w_n(B) \leq w_{n-1}(B) \quad \text{(the isoperimetric inequality)}.$$

See Chapter 1 for a different approach to these two inequalities.

A fortiori, the inequalities of Alexandrov imply that for any $k \leq n$ we have

$$\left(\frac{\text{vol}(B)}{V_n}\right)^{1/n} \leq \sup\left\{\left(\frac{\text{vol}(P(B))}{V_k}\right)^{1/k}\middle| rk(P) = k\right\},$$

which immediately implies, for all $K \subset H$ as above,

$$v_n(K) \leq v_k(K) \text{ if } k \leq n.$$

Actually, the isoperimetric inequality (cf. Corollary 1.3) suffices to prove this since it gives us $w_n(B) \leq w_{n-1}(B)$ for any $B \subset \mathbf{R}^n$, hence $v_n(K) \leq v_{n-1}(K)$ for any $K \subset H$. The numbers $w_k(B)$ are especially useful via the following classical Steiner–Minkowski formula (cf. again [Al], [BF], [BZ], or [Eg]):

$$\forall\, t > 0 \quad \frac{\text{vol}(B + tB_{\ell_2^n})}{V_n} = \sum_{0 \leq k \leq n} \binom{n}{k} t^{n-k} w_k(B)^k.$$

However, we will not use this beautiful formula in the sequel.

We will denote by $\mathcal{P}$ the set of all orthogonal projections on H and by $\mathcal{P}_k$ the set of all orthogonal projections of rank k. By *projection*, we always mean in this chapter an orthogonal projection.

We now compare $v_n(T)$ and $e_n(T)$. First note that if P is in $\mathcal{P}_n$, we have

$$\text{vol}(P(\overline{T(B_X)})) \leq 2^{n-1}e_n(PT)^n V_n;$$

hence

$$v_n(T) \leq 2\, e_n(T).$$

In this chapter, we seek results in the converse direction: we will assume a majorization of $v_n(T)$ and will deduce a similar bound for $e_n(T)$. We may formulate the main result of this chapter as follows.

Theorem 9.1. *Let X be a Banach space.*

(i) *There is a numerical constant C such that for all Hilbert spaces H and all operators $u : H \rightarrow X$ we have*

(9.2) $$\ell(u) \leq C \sum_{n\geq 1} v_n(u^*)n^{-1/2}(1 + \text{Log } n).$$

(ii) *More precisely, for each $k \geq 1$ let*

$$\ell_k(u) = \inf\{\ell(u - uP)\},$$

where the infimum runs over all orthogonal projections P on H with rank $P < k$. *There are numerical constant $C' > 0$ and $C'' > 0$ such that for all Hilbert spaces H and all $u : H \rightarrow X$ we have for each $k \geq 1$*

(9.3) $$\ell_k(u) \leq C'' \sum_{n\geq C'K} v_n(u^*)n^{-1/2}(1 + \text{Log } n).$$

In particular, if $\sum_{n\geq 1} v_n(u^*)n^{-1/2}(1 + \text{Log } n) < \infty$ then $\ell(u) < \infty$ and u belongs to the closure of the finite-rank operators in the space $\ell(H, X)$.

(iii) *Finally, if X is K-convex, the preceding statements hold without the factor $(1 + \text{Log } n)$.*

The crucial lemma in the proof may be formulated like this.

Lemma 9.2. *There is a numerical constant C_1 such that the following holds. Let n be an arbitrary integer and let B be any ball in $\mathbf{R}^n$. Let $k < m < n$. There is a subspace $F \subset \mathbf{R}^n$ with* $\dim F = m - k + 1$ *such that*

$$\text{diam}(F \cap B) \leq C_1 v_k(B)f\left(\frac{m}{n}\right),$$

where $f(\delta) = (1 - \delta)^{-1}[1 + \text{Log }(1 - \delta)^{-1}]$.

Proof: We first consider the particular case where B is an ellipsoid. Then we may assume (without loss of generality) that there are $\lambda_1 \geq \lambda_2 \ldots \geq \lambda_n > 0$ such that

$$B = \left\{ x \in \mathbf{R}^n \,\middle|\, \sum_1^n (X_i/\lambda_i)^2 \leq 1 \right\}.$$

Then clearly

$$\left| \prod_{i=1}^k \lambda_i \right| \leq v_k(B)^k,$$

so that $\lambda_k \leq v_k(B)$.

Now, if we let F be the span of $e_k, e_{k+1}, \ldots, e_n$ we find $\dim F = n - k + 1$ and $\operatorname{diam}(B \cap F) \leq v_k(B)$.

We now turn to the general case. We will use the QS-Theorem proved in the preceding chapter. Let $k < m < n$. By the Remark after Theorem 8.4., we know that there is an m-dimensional projection of a section of B which is $C_1 f(m/n)$-equivalent to an ellipsoid D, where C_1 is a numerical constant. Hence there are $E_2 \subset E_1 \subset \mathbf{R}^n$ and $D \subset E_2$ such that

(9.4) $$D \subset P_{E_2}(E_1 \cap B) \subset C_1 f\left(\frac{m}{n}\right) D.$$

Now we will compare $v_k(D)$ and $v_k(B)$. Clearly,

$$P_{E_2}(E_1 \cap B) \subset P_{E_2}(B);$$

hence

$$v_k(P_{E_2}(E_1 \cap B)) \leq v_k(P_{E_2}(B)) \leq v_k(B)$$

where the second inequality follows from definition (9.1).

By (9.4), this implies

(9.5) $$v_k(D) \leq v_k(B).$$

By the first part of the proof, there is a subspace $F \subset E_2$ with $\dim F = m - k + 1$ such that

$$\operatorname{diam}(F \cap D) \leq v_k(D);$$

hence by (9.4) and (9.5)

$$\operatorname{diam}(F \cap B) \leq C_1 f\left(\frac{m}{n}\right) v_k(D) \leq C_1 f\left(\frac{m}{n}\right) v_k(B). \quad \blacksquare$$

As a typical application, we can state (take $k = \left[\frac{n}{4}\right]$ and $m = \left[\frac{3n}{4}\right]$).

Corollary 9.3. *There is a numerical constant $C_2 > 0$ such that, for all n, for any ball $B \subset \mathbf{R}^n$ there is a subspace $F \subset \mathbf{R}^n$ of dimension $\left[\frac{n}{2}\right]$ such that*

$$\text{diam}(F \cap B) \le C_2 v_{\left[\frac{n}{4}\right]}(B).$$

We will also need the following lemmas which are related to the bounds for the K-convexity constant presented in Chapter 2.

Lemma 9.4. *Let S be a closed subspace of a Banach space E. Let $\sigma : E \to E/S$ be the quotient map. Assume that E is K-convex. Let x_n in E/S be such that $\sum_{n=1}^{\infty} g_n x_n$ converges in $L_2(E/S)$. Then for each $\epsilon > 0$ there are $\tilde{x}_n$ in E such that $\sigma(\tilde{x}_n) = x_n$ and the series $\sum_{n=1}^{\infty} g_n \tilde{x}_n$ converges in $L_2(E)$ to a limit satisfying*

$$\left\| \sum_1^{\infty} g_n \tilde{x}_n \right\|_{L_2(E)} \le K(X)(1 + \epsilon) \left\| \sum_1^{\infty} g_n x_n \right\|_{L_2(E/S)}.$$

Proof: This clearly reduces to the case of a *finite* sequence $(x_1, \ldots, x_n)$. Then, let $\phi = \sum_1^n g_i x_i$. We consider ϕ as an element of $L_2(E/S)$, or equivalently $L_2(E)/L_2(S)$. Clearly, there is $\tilde{\phi}$ in $L_2(E)$ such that $\sigma \tilde{\phi} = \phi$ and

(9.6) $$\|\tilde{\phi}\|_{L_2(E)} \le (1 + \epsilon)\|\phi\|_{L_2(E/S)}.$$

Moreover, we may clearly assume that $\tilde{\phi}$ depends only on $g_1, \ldots, g_n$. With the notation of Chapter 2, let

$$\psi = (Q_1 \otimes I_E)(\tilde{\phi}).$$

Then ψ admits an *a priori* expansion

$$\psi = \sum_1^n g_i \tilde{x}_i \text{ with } \tilde{x}_i \in E.$$

Obviously, $\sigma \psi = \phi$, hence we must have $\sigma \tilde{x}_i = x_i$ for all i; and on the other hand, by definition,

$$\|\psi\|_{L_2(E)} \le K(E)\|\tilde{\phi}\|_{L_2(E)};$$

and by (9.6),

$$\le K(E)(1 + \epsilon)\|\phi\|_{L_2(E/S)}. \quad \blacksquare$$

Remark 9.5: Recall that for an arbitrary n-dimensional normed space E we have proved in Chapter 2 that $K(E) \leq K(1 + \text{Log } n)$. Thus, Lemma 9.4 can be useful also in that case, as we shall see in the proof of Theorem 9.1. Note also that Lemma 9.4 may be reformulated as follows: For each $\epsilon > 0$, every operator $v : \ell_2^n \to E/S$ admits a lifting $\widetilde{v} : \ell_2^n \to E$, satisfying $\sigma \widetilde{v} = v$ and $\ell(\widetilde{v}) \leq K(E)(1 + \epsilon)\ell(v)$.

We need one more lemma.

Lemma 9.6.   *Let n be an integer. Let $u : \ell_2^{4n} \to X$ be an operator into a Banach space X. Then there is an orthogonal projection P on ℓ_2^{4n} with rank $2n$ such that*

$$\ell(uP) \leq C_3 v_n(u^*)n^{1/2}(1 + \text{Log } n),$$

where $C_3 > 0$ is a numerical constant.

Proof: We may clearly assume $\dim X \leq 4n$. By Corollary 9.3 we know that there is a subspace $F \subset \mathbf{R}^{4n}$ with dimension $2n$ such that

$$\|u^*_{|u^{*-1}(F)}\| \leq C_2 v_n(u^*).$$

Let $F_1 = u^{*-1}(F) \subset X^*$. Then $F_1^{\perp} = u(F) \subset X$. Let $\sigma : X \to X/F_1^{\perp}$ be the quotient map. Then $\|\sigma u\| = \|u^*_{|F_1}\| \leq C_2 v_n(u^*)$.

We have, trivially,

$$\ell(\sigma u) \leq (4n)^{1/2}\|\sigma u\|$$
$$\leq (4n)^{1/2}C_2 v_n(u^*).$$

By Lemma 9.4 and Remark 9.5, there is an operator $\widetilde{u} : \ell_2^{4n} \to X$ such that $\sigma \widetilde{u} = \sigma u$ and

$$\ell(\widetilde{u}) \leq 2K(1 + \text{Log } 4n)\ell(\sigma u).$$

Hence

$$\ell(\widetilde{u}) \leq C_3 v_n(u^*)n^{1/2}(1 + \text{Log } n)$$

for some numerical constant $C_3 > 0$.

Finally, we note that $\sigma(\widetilde{u} - u) = 0$ implies that the image of $\widetilde{u} - u$ lies in $u(F)$, hence $rk(\widetilde{u} - u) \leq \dim F \leq 2n$, which implies $\dim \text{Ker}(\widetilde{u} - u) \geq 2n$. Let P be the orthogonal projection onto $\text{Ker}(\widetilde{u} - u)$. We have $(\widetilde{u} - u)P = 0$, hence

$$\ell(uP) = \ell(\widetilde{u}P) \leq C_3 v_n(u^*)n^{1/2}(1 + \text{Log } n). \quad \blacksquare$$

Proof of Theorem 9.1: Consider an operator $u : H \to X$. We will first prove (9.2). Let $\lambda(2^n) = \sup\{\ell(uQ)|Q \in \mathcal{P}_{2^n}\}$. Note that $\ell(u) = \sup_{n \geq 0} \lambda(2^n)$. By Lemma 9.6, for any projection Q in $\mathcal{P}_{2^n}$ there is a projection P in $\mathcal{P}_{2^{n-1}}$ with $P \leq Q$ such that

$$\ell(uP) \leq C_3 v_{2^{n-2}}(Qu^*)(2^{n-2})^{1/2}(1 + \text{Log } 2^{n-2})$$
$$\leq C_3 v_{2^{n-2}}(u^*)2^{n/2}(1 + \text{Log } 2^n).$$

Clearly, $\ell(uQ) \leq \ell(uP) + \ell(u(Q - P))$; hence, since $Q - P \in \mathcal{P}_{2^{n-1}}$,

$$\ell(uQ) \leq C_3 v_{2^{n-2}}(u^*)2^{n/2}(1 + \text{Log } 2^n) + \lambda(2^{n-1}).$$

Thus we have proved

$$\lambda(2^n) \leq C_3 v_{2^{n-2}}(u^*)2^{n/2}(1 + \text{Log } 2^n) + \lambda(2^{n-1}),$$

which implies

$$\lambda(2^n) \leq C_3 \sum_{n \geq 2} v_{2^{n-2}}(u^*)2^{n/2}(1 + \text{Log } 2^n) + \lambda(2).$$

Clearly, $\lambda(2) \leq 2\|u\| = 2v_1(u^*)$, so that we can write

$$\ell(u) = \sup \lambda(2^n) \leq C_4 \sum_{k \geq 0} v_{2^k}(u^*)2^{k/2}(1 + \text{Log } 2^k)$$
$$\leq C_5 \sum_{n \geq 1} v_n(u^*)n^{-1/2}(1 + \text{Log } n)$$

for some numerical constants C_4, C_5. This proves (9.2).

Thus we have proved that

(9.7)
$$\sum_{n \geq 1} v_n(u^*)n^{-1/2}(1 + \text{Log } n) < \infty$$

implies $\ell(u) < \infty$. By the results of Chapter 5 (cf. Theorem 5.5), it follows that u is compact. Hence, for each $\epsilon > 0$, there is a projection Q in $\mathcal{P}$ such that $\|u - uQ\| < \epsilon$.

Obviously this implies $v_n((u - uQ)^*) \leq \epsilon$. On the other hand, clearly

$$v_n((u - uQ)^*) \leq v_n(u^*),$$

hence

$$v_n((u - uQ)^*) \leq \min\{\epsilon, v_n(u^*)\}.$$

By (9.2), this shows that

(9.8) $\ell(u - uQ) \leq C\beta(\epsilon)$

where

$$\beta(\epsilon) = \sum_{n \geq 1} \min\{\epsilon, v_n(u^*)\} n^{-1/2}(1 + \text{Log } n).$$

Obviously (9.7) implies $\beta(\epsilon) \to 0$ when $\epsilon \to 0$.

For simplicity, we may (and do) assume without loss of generality that H is infinite dimensional and that $rk(Q) = 2^K$ where K is an integer depending on ϵ. Then we can view uQ essentially as an operator from $\ell_2^{2^k}$ into X to which we apply Lemma 9.6. This shows that there exists P_1 in $\mathcal{P}_{2^{K-1}}$ such that $P_1 \leq Q$ and

$$\ell(uP_1) \leq C_3 v_{2^{K-2}}(u^*)(2^{K-2})^{1/2}(1 + \text{Log } 2^{K-2}).$$

Applying the same reasoning with $Q - P_1$ instead of Q, we obtain $P_2 \leq Q - P_1$, applying it one more time with $Q - P_1 - P_2$ instead of $Q - P_1$, we obtain $P_3 \leq Q - P_1 - P_2$, and so on. This reasoning yields a sequence of projections $\{P_j\}$ such that

(9.9)
$$rk(Q) - rk(P_1 + \ldots + P_j) = 2^{K-j},$$
$$P_j \leq Q - (P_1 + \ldots + P_{j-1})$$

and

(9.10) $\ell(uP_j) \leq C_3 v_{2^{K-j-1}}(u^*)(2^{k-j-1})^{1/2}(1 + \text{Log } 2^{K-j-1}).$

Now let $J = K - m$ for some $m \leq K$. We define

$$\widetilde{Q} = Q - (P_1 + \ldots + P_J).$$

We have $rk(Q - \widetilde{Q}) = 2^m$ by (9.9) and by (9.10)

$$\ell(uQ - u\widetilde{Q}) \leq \ell(uP_1 + \ldots + uP_J)$$
$$\leq C_3 \sum_{1 \leq j \leq J} C_3 v_{2^{K-j-1}}(u^*)(2^{K-j-1})^{1/2}(1 + \text{Log } 2^{K-j-1}).$$

Therefore we have

$$\ell(uQ - u\widetilde{Q}) \leq C_3 \alpha_m,$$

where

$$\alpha_m = \sum_{i \geq m-1} v_{2^i}(u^*) 2^{i/2} (1 + \text{Log } 2^i).$$

Recalling (9.8), we obtain

$$\ell(u - u\widetilde{Q}) \leq \beta(\epsilon) + C_3 \alpha_m.$$

Since $rk\widetilde{Q} = 2^m$, we find

$$\ell_{2^m+1}(u) \leq \beta(\epsilon) + C_3 \alpha_m$$

and since $\epsilon > 0$ is arbitrary, this yields

$$\ell_{2^m+1}(u) \leq C_3 \alpha_m \text{ for all } m \geq 1.$$

It is then an elementary task to check that there are positive constants C', C'' such that (9.3) holds. This proves the second part of Theorem 9.1.

Finally, the third part is clear since, if X is K-convex, we may replace all the $(1 + \text{Log } n)$ factors in the above proof by $K(X)$. ∎

We have seen above that $v_n(u^*) \leq 2e_n(u^*)$. Thus $e_n(u^*) \in 0(n^{-\alpha}) \Rightarrow v_n(u^*) \in 0(n^{-\alpha})$.

In the converse direction, we can prove only a slightly weaker fact, as follows.

Corollary 9.7. *Consider $u : H \to X$ as above. Then we have*

$$\limsup_{n \to \infty} \frac{\text{Log } e_n(u^*)}{\text{Log } n} = \limsup_{n \to \infty} \frac{\text{Log } v_n(u^*)}{\text{Log } n},$$

provided the right-hand side is $< -1/2$

Proof: By the preceding remark, it suffices to show that if $v_n(u^*)$ is $0(n^\alpha)$ with $\alpha < 1/2$ then for all $\delta > 0$ $e_n(u^*)$ is $0(n^{\alpha+\delta})$.

Actually, we will establish the following more precise claim: there are constants $\delta > 0$ and C_6 such that for all $k \geq 1$ we have

$$c_k(u^*) \leq C_6 k^{-1/2} \sum_{n \geq \delta k} v_n(u^*) n^{-1/2} (1 + \text{Log } n).$$

This follows from (9.3) and the results of Chapter 5.

Indeed, by (9.3), for each $k \geq 1$ there is a projection P of rank $< k$ such that

$$\ell(u - uP) \leq 2C'' \sum_{n \geq C'k} v_n(u^*)n^{-1/2}(1 + \text{Log } n).$$

Hence, by Theorem 5.8,

$$c_k((1 - P)u^*) \leq C_7 \sum_{n \geq C'k} v_n(u^*)n^{-1/2}(1 + \text{Log } n).$$

Clearly, $c_{2k}(u^*) \leq c_k((1 - P)u^*)$ since $rk(P) < k$, therefore we can easily conclude that the above claim is valid for suitable constants $\delta > 0$ and C_6.

From this claim, we deduce that if $v_n(u^*)$ is $0(n^\alpha)$ with $\alpha < -1/2$, then $c_n(u^*)$ is $0(n^\alpha \text{ Log } n)$, hence $c_n(u^*)$ is $0(n^{\alpha+\delta})$ for all $\delta > 0$, and by Theorem 5.2 this implies *a fortiori* that $e_n(u^*)$ is $0(n^{\alpha+\delta})$. ∎

Notes and Remarks

This chapter is mainly based on [MP2], but our exposition is different. We follow the approach suggested in the note *added in proof* to [MP2]. I am indebted to A. Pajor for several conversations which considerably clarified the presentation of this chapter. Except for Lemma 9.4, which is a basic property of K-convex spaces (see [P6] for a more refined result), all the results come from [MP2].

The motivation for this chapter came from the 1967 paper of Dudley [Du1] where he connected the *geometric* study of compact subsets of Hilbert space with the continuity of the paths of Gaussian processes. We can describe briefly Dudley's viewpoint as follows. Let H be a (separable) Hilbert space and let $(X_t)_{t \in H}$ be a Gaussian process indexed by H such that

$$(9.11) \qquad \forall \, t, s \in H \quad \|X_t - X_s\|_2 = \|t - s\|_H.$$

A subset $K \subset H$ is called a GC-set (resp. GB-set) if the process $(X_t)_{t \in K}$ has a version which is continuous on K (resp. bounded on K).

A process $(\widetilde{X}_t)$ is called a version of X_t if we have each $\widetilde{X}_t \overset{\text{a.s.}}{=} X_t$ for t. Since the marginal distributions of Gaussian processes are determined by their covariance, it is easy to see that these definitions do not depend on the particular choice of the process (X_t) as long as it satisfies (9.11). It is well known that there exists a Gaussian process satisfying (9.11) and such that $t \to X_t$ is a linear map from H into $L_2(\Omega, \mathcal{A}, \mathbf{P})$. To see this, just take $H = \ell_2$ and let

$$(9.12) \qquad X_t = \sum t_n g_n,$$

with (g_n) i.i.d. normal Gaussian variables as usual.

Actually, the same argument (consider the marginal distributions of X_t restricted to K) shows that if there is a (not necessarily linear) isometry from a set K_1 onto a set K_2 then K_1 is a GC-set (resp. GB-set) iff K_2 is also one. More generally, it was later proved by Sudakov [Su] that if there is a (non-linear) contraction from K_1 onto K_2 and if K_1 is a GC-set (resp. GB-set) then K_2 is also one.

In his fundamental 1967 paper, Dudley proved the sufficiency of the metric entropy condition as proved in Chapter 5. He also proved that the condition $\sup_n n^{1/2} v_n(K) < \infty$ is *necessary* for K to be a GB-set and in the converse direction he conjectured that $\sup_n n^{1/2+\epsilon} v_n(K) < \infty$ for some $\epsilon > 0$ is sufficient for K to be a GC-set. This was first proved in [MP2]. (It clearly follows from Corollary 9.7, which was also conjectured by Dudley.) The necessity of $\sup_n n^{1/2} v_n(K)$ can be proved (essentially as Dudley) using the Urysohn inequality (cf. Corollary 1.4). Indeed, we have

$$v_n(K) \le \int_S \sup_{t \in K} <t, x> \, d\sigma(x).$$

But, with (X_t) as in (9.12),

$$\int \sup_{t \in K} | <t, x> |^2 \, d\sigma(x) = n^{-1} \mathbf{E} \sup_{t \in K} |X_t|^2;$$

hence for all $K \subset H$ we have

$$v_n(K) \le n^{-1/2} \sup_{P \in \mathcal{P}_n} \left(\mathbf{E} \sup_{t \in PK} |X_t|^2 \right)^{1/2}$$

$$\le n^{-1/2} \left(\mathbf{E} \sup_{t \in K} |X_t|^2 \right)^{1/2}.$$

Finally, we may write for all $u : \ell_2 \to X$

$$v_n(u^*) \le n^{-1/2} \ell(u).$$

(Recall also that $(v_n(u^*) \le 2e_n(u^*)$, as noted in the beginning of Chapter 9, so that this can be viewed also as a consequence of the lower bound in Theorem 5.5.)

We refer to [PT2] and [PT4] for more results using volume numbers.

We also refer to [Pa2] and [Pa3] for more information on the mixed volumes when the Euclidean unit ball is replaced by the unit ball of ℓ_∞^n or of ℓ_1^n. In particular, Pajor proved a version of Urysohn's inequality (cf. Corollary 1.4) for random choices of signs. Namely, he proved that for all compact subsets $K \subset \mathbf{R}^n$ we have

$$\left(\frac{\text{vol}(K)}{\text{vol}(B_{\ell_1^n})} \right)^{1/n} \leq \int \sup_{t \in K} \langle t, x \rangle \, d\mu(x),$$

where μ is the uniform probability measure on $\{-1, 1\}^n$. See also [El] and [Pa2] for related information.

Chapter 10

Weak Cotype 2

This chapter is mainly based on [MP1], where the notion of weak cotype 2 was first introduced and developed. The previous paper [BM] contained a result on cotype 2 spaces, which considerably influenced the motivation for the paper [MP1].

To explain the notation of weak cotype 2, we first define the more usual notion of cotype 2. Let $2 \leq q < \infty$. Let X be a Banach space. We say that X is a cotype q space if there is a constant C such that for all n and for all $x_1, \ldots, x_n$ in X we have

$$\left(\sum \|x_k\|^q\right)^{1/q} \leq C \left\|\sum_1^n g_k x_k\right\|_{L_2(X)}.$$

The smallest constant C for which this holds is called the cotype q constant of X and is denoted by $C_q(X)$. We refer to [MaP], [P1], and [MS] for more information and examples.

The notion of cotype q is usually defined using a sequence (ε_n) of independent, symmetric, ± 1-valued random variables instead of the sequence (g_n). It is well known, however, that the two definitions are equivalent (cf. [MaP], see also e.g. [P1], p. 187).

Consider now an operator $u : \ell_2^n \to X$. By Lemma 1.8 there is an orthonormal basis $f_1, \ldots, f_n$ in ℓ_2^n such that $a_k(u) \leq \|u(f_k)\|$ for $k = 1, 2, \ldots, n$. Moreover, we have clearly (by the rotational invariance of Gaussian measures, cf. (2.1) above)

$$\ell(u) = \left\|\sum_{k=1}^n g_k u(f_k)\right\|_{L_2(X)}.$$

Therefore, if X is of cotype q, we have necessarily (taking $x_k = u(f_k)$)

$$\left(\sum_{k=1}^n a_k(u)^q\right)^{1/q} \leq C_q(X)\ell(u)$$

and *a fortiori* we find

$$\sup_{k \geq 1} k^{1/q} a_k(u) \leq C_q(X)\ell(u).$$

Now we introduce weak cotype 2.

Definition 10.1.   *Let X be a Banach space. We say that X is a weak cotype 2 space if there is a constant C such that, for all n and all operators $u : \ell_2^n \to X$, we have*

$$(10.1) \qquad\qquad \sup_{k \geq 1} k^{1/2} a_k(u) \leq C\, \ell(u).$$

The smallest constant C for which this holds will be denoted by $wC_2(X)$.

Clearly, the above preliminary remarks show that cotype 2 implies weak cotype 2 and $wC_2(X) \leq C_2(X)$. Of course, (10.1) immediately extends to all operators $u : \ell_2 \to X$ for which $\ell(u) < \infty$.

The notion of weak cotype 2 is clearly inherited by the subspaces of a Banach space (but not by the quotient spaces in general; see below). Actually, it is stable by *finite representability*. Recall that a Banach space Y is said to be finitely representable into X (in short, Y f.r. X) if for every $\varepsilon > 0$ and every f.d. subspace $E \subset Y$ there is a subspace $F \subset X$ such that $d(E, F) < 1 + \varepsilon$.

It is easy to see that Y f.r. X and X of weak cotype 2 implies Y weak cotype 2 also. (Thus weak cotype 2 is a super-property in the sense of James [J].) Clearly, the direct sum of two (or a finite number of) weak cotype 2 spaces is again a weak cotype 2 space.

We will show below that weak cotype 2 characterizes the class of Banach spaces for which the dimension of the spherical sections of the unit ball (as in Dvoretzky's Theorem) is always essentially the largest possible. More precisely, to measure these dimensions and their asymptotic behavior, we introduce the following (possibly infinite) number for $0 < \delta < 1$:

$$d_X(\delta) = \sup_{\substack{E \subset X \\ E \text{ f.d.}}} \inf\{d_F \,|\, F \subset E \ \dim F \geq \delta \ \dim E\}.$$

In other words, $d_X(\delta)$ is the smallest constant C such that every f.d. subspace $E \subset X$ contains a subspace $F \subset E$ with $\dim F \leq \delta \dim E$ such that $d_F \leq C$.

Clearly, if X is a Hilbert space, then $d_X(\delta) = 1$ for all $0 < \delta < 1$. In general, of course, $d_X(\delta)$ may be infinite.

We first connect the notion of weak cotype 2 with the existence of Euclidean subspaces of proportional dimension in every $E \subset X$, as follows.

Theorem 10.2. *Let X be a Banach space. The following properties are equivalent.*

(i) *X is a weak cotype 2 space.*
(ii) *There is $0 < \delta_o < 1$ such that $d_X(\delta_o)$ is finite.*
(iii) *For each $0 < \delta < 1, d_X(\delta)$ is finite.*
(iv) *There is a constant C such that for all $0 < \delta < 1$ we have*

$$d_X(\delta) \le C(1 - \delta)^{-1}\left[1 + \text{Log}\,\frac{C}{1 - \delta}\right].$$

Note: In this chapter the properties numbered from (i) to (viii) will all be equivalent to weak cotype 2.

Remark: The proof below of (ii) $\Rightarrow$ (iv) gives an estimate of $C \le C'd_X(\delta_0)\delta_o^{-1}$ for some numerical constant C'.

For the proof of Theorem 10.2, we will need the following:

Lemma 10.3. *Let $u : H \to X$ be an operator on a Hilbert space H with values in a Banach space X.*

Let $0 < \alpha < 1, \beta > 0$ and $\lambda > 0$ such that $\alpha \dim H \ge 1$. Assume that for any n and any n-dimensional subspace $S \subset H$ we have (denoting [ ] the integral part)

$$(10.2) \qquad a_{[\alpha n]}(u_{|S}) \le \lambda[\alpha n]^{-\beta} \quad \text{whenever } [\alpha n] \ge 1;$$

then we have

$$(10.3) \qquad \sup_{k \ge 1} k^\beta a_k(u) \le \lambda C \quad \text{with } C = 3^\beta(1 - \alpha^{2\beta})^{-1/2}.$$

Proof: We may clearly assume $\lambda = 1$ by homogeneity. Let C_n be the best constant C for which the preceding statement holds for all $u : H \to X$ with rank $\le n$. Clearly $C_n < \infty$. We will show by an *a priori* reasoning that C_n remains bounded when $n \to \infty$. More precisely, we claim that

$$C_n \le 3^\beta(1 - \alpha^{2\beta})^{-1/2}.$$

To prove this claim, we consider $u : H \to X$ with rank $\le n$. Let $m = [k/\alpha]$. By definition of $a_{m+1}(u)$ there is a subspace $S \subset H$ such that

$$(10.4) \qquad \|u_{|S}\| = a_{m+1}(u) \le C_n(m + 1)^{-\beta}.$$

We may identify $S^\perp$ with ℓ_2^m and apply the assumption of Lemma 10.3 to $u_{|S^\perp}$. Thus we find

$$(10.5) \qquad a_{[\alpha m]}(u_{|S^\perp}) \le [\alpha m]^{-\beta}.$$

Let P_S and $P_{S^\perp}$ be the orthogonal projections onto S and $S^\perp$ respectively. Note that for any $w : H \to X$ we have

$$\|w\| \le (\|wP_S\|^2 + \|wP_{S^\perp}\|^2)^{1/2}.$$

Thus, (10.4) and (10.5) imply

$$(10.6) \qquad a_{[\alpha m]}(u) \le (C_n^2 (m+1)^{-2\beta} + [\alpha m]^{-2\beta})^{1/2},$$

provided, of course, $[\alpha m] \ge 1$ (which is guaranteed if $k \ge 2$). Since $k \ge \alpha m$, we have $a_k(u) \le a_{[\alpha m]}(u)$ and, moreover, $m + 1 \ge k/\alpha$ and $[\alpha m] \ge k - \alpha - 1$ so that (10.6) implies, for all $k > 2$,

$$(10.7) \qquad \begin{aligned} k^{2\beta} a_k(u)^2 &\le C_n^2 \alpha^{2\beta} + (k(k - \alpha - 1)^{-1})^{2\beta} \\ &\le C_n^2 \alpha^{2\beta} + 3^{2\beta}. \end{aligned}$$

On the other hand, there is obviously an integer $n_o = n_o(\alpha)$ such that $[\alpha n_o] = 1$, hence (10.2) implies $\|u\| \le 1$. This observation and (10.7) imply

$$\sup_{k \ge 1} k^{2\beta} a_k(u)^2 \le \alpha^{2\beta} C_n^2 + 3^{2\beta}.$$

Therefore, we have (going back to the definition of C_n)

$$C_n^2 \le \alpha^{2\beta} C_n^2 + 3^{2\beta};$$

hence (since $\alpha^{2\beta} < 1$) $C_n \le 3^\beta (1 - \alpha^{2\beta})^{-1/2}$, as claimed earlier. It is then easy to deduce that any operator u satisfying (10.2) must be compact and must satisfy (10.3). ∎

Proof of Theorem 10.2: We will show that (iv) ⇒ (iii) ⇒ (ii) ⇒ (i) ⇒ (iv). The implications (iv) ⇒ (iii) ⇒ (ii) are trivial. Let us show that (ii) ⇒ (i). Assume (ii). Let H be a Hilbert space and consider $u : H \to X$ with $\ell(u) = \sup\{\ell(u_{|S}), S \subset H\} < \infty$. We will prove that (10.1) holds using Lemma 10.3. Let $\alpha = 1 - \delta_0/2$. Consider a subspace $S \subset H$ with $n = \dim S$. We claim that, if $[\alpha n] \ge 1$, we have

$$(10.8) \qquad a_{[\alpha n]}(u_{|S}) \le C' \ell(u)[\alpha n]^{-1/2}$$

for some constant C'.

Indeed, we may assume (by a perturbation argument) that $u_{|S}$ is an isomorphism. Let then $E = u(S)$. There is a subspace $F \subset E$ with $\dim F \geq \delta_0 n$ and $d_F \leq C$ with $C = d_X(\delta_0)$. Let $S_1 = u^{-1}(F)$. The operator $u_{|S_1} : S_1 \to F$ may be viewed as an operator between Hilbert spaces up to the constant C, hence in particular we have, by Proposition 3.13,

$$\left(\sum a_k(u_{|S_1})^2 \right)^{1/2} \leq C\ell(u_{|S_1})$$
$$\leq C\ell(u).$$

This implies $a_k(u_{|S_1}) \leq C\ell(u)k^{-1/2}$ for all $k \geq 1$. Let $m = n - \dim S_1$. This implies $a_{m+k}(u) \leq C\ell(u)k^{1/2}$ for all $k \geq 1$. Choosing $k = [\frac{\delta_0 n}{2}]$ we find $m + k \leq [\alpha n]$ and (10.8) follows with $C' \leq C'' d_X(\delta_0)/\sqrt{\delta_0}$ for some numerical constant C'', if we assume $[\frac{\delta_0 n}{2}] \geq 1$. On the other hand, the case $[\frac{\delta_0 n}{2}] < 1$ is trivial. Indeed, if $n < 2/\delta_0$ then obviously

$$\sup_{k \leq n} k^{1/2} a_k(u) \leq n^{1/2} \|u\| \leq (2/\delta_0)^{1/2} \ell(u).$$

Thus, we may conclude that (10.8) holds, and, by Lemma 10.3, X is a weak cotype 2 space with $wC_2(X) \leq C_3 d_X(\delta_0)(\delta_0)^{-1}$ for some numerical constant C_3. This concludes the proof that (ii) $\Rightarrow$ (i).

Let us now show that (i) $\Rightarrow$ (iv). We will give a proof which uses the iteration argument described in Chapter 8. (For a proof which does not use this, see [MP1].) Let X be a weak cotype 2 space. Let E be a f.d. subspace of X with $\dim E = n$. By Theorem 3.11, there is an isomorphism $u : \ell_2^n \to E$ satisfying $\ell(u) = n^{1/2}$ and $\ell(u^{-1*}) \leq K(1 + \text{Log } d_E)n^{1/2}$. By Theorem 5.8, there is a subspace $F_1 \subset E$ with $\text{codim } F_1 < k$ such that

$$(10.9) \qquad \|u_{|F_1}^{-1}\| \leq K(1 + \text{Log } d_E)(n/k)^{1/2}.$$

Consider $S_1 = u^{-1}(F_1)$ and $u_{|S_1} : S_1 \to E$. By (10.1) there is a subspace $S_2 \subset S_1$ with $\dim S_1/S_2 < k$ such that

$$(10.10) \qquad \begin{aligned} \|u_{|S_2}\| &\leq wC_2(X)\ell(u_{|S_1})k^{-1/2} \\ &\leq wC_2(X)(n/k)^{1/2}. \end{aligned}$$

Let $F = u(S_2)$. Then we have $\dim F > n - 2k + 1, F \subset F_1$, and, by (10.9) and (10.10),

$$\begin{aligned} d_F &\leq \|u_{|S_2}\| \cdot \|u_{|F}^{-1}\| \\ &\leq wC_2(X)(n/k)^{1/2} K(1 + \text{Log } d_E)(n/k)^{1/2}. \end{aligned}$$

By an obvious adjustment of the constant K, we can replace $2k$ by k in
the preceding estimate and we obtain for every $k = 1, \ldots, n$ a subspace
$F \subset E$ with $\dim F = n-k$ such that $d_F \leq K w C_2(X)(n/k)(1+\mathrm{Log}\, d_E)$
where K is a numerical constant.

By the iteration argument of Chapter 8 (cf. Lemma 8.6), for every
$k = 1, \ldots n$, there is a subspace $F \subset E$ with $\dim F = n - k$ such that,
for every $k = 1, \ldots, n$, there is a subspace $F \subset E$ with $\dim F = n - k$
such that

$$d_F \leq K' w C_2(X)\left(\frac{n}{k}\right)\left(1 + \mathrm{Log}\left(K' w C_2(X)\frac{n}{k}\right)\right),$$

where K' is a numerical constant.

This concludes the proof of that (i) $\Rightarrow$ (iv) and the proof Theorem
10.2 is complete.

Remark: In case X is of cotype 2, it is possible to prove a better
estimate then (iv) above, namely

$$d_X(\delta) \leq K(1 - \delta)^{-1/2}(1 + \log(1 - \delta)^{-1})$$

for some constant K. This was first proved in [PT1] (cf. [MP1] for a
slightly more direct proof). We do not know whether such an estimate
is valid in a weak cotype 2 space.

We now come to the relationship with the notion of volume ratio
introduced in Chapter 6.

Theorem 10.4.   *Let X be a Banach space. The following are equivalent.*

  (i) *X is a weak cotype 2 space.*
  (v) *There is a constant C such that for all n, for all n-dimensional
      subspaces $E \subset X$ and for all $v : E \to \ell_2^n$ we have*

$$e_n(v) \leq C\ \pi_2(v)n^{-1/2}.$$

 (vi) *There is a constant C such that for all f.d. subspaces $E \subset X$ we
      have $vr(E) \leq C$.*

Remark: The constants involved in the preceding equivalent properties can all be estimated in the usual manner. For instance, in the proof of (i) $\Rightarrow$ (vi) below, we will actually establish the following more precise statement: There is a function $\lambda \to f(\lambda)$ such that any Banach space X with $w C_2(X) \leq \lambda$ must satisfy (vi) with $C \leq f(\lambda)$.

For the proof we will use the following simple result.

Lemma 10.5. *Let E be an n-dimensional space. Assume that there is a constant D and $\alpha > 0$ such that for all $k \leq n$ there is a subspace $F \subset E$ with $\dim F > n - k$ such that $d_F \leq D(n/k)^\alpha$. Let $v : E \to \ell_2^n$ be an operator. Then we have for all $k*

$$c_k(v) \leq D(n/k)^\alpha \pi_2(v) k^{-1/2}.$$

Proof: Let j be an integer with $1 \leq j \leq n$. Let $F \subset E$ satisfy $d_F \leq C(n/j)^\alpha$ and $\dim F > n - j$. We may consider $v_{|F} : F \to \ell_2^n$ as an operator between Hilbert spaces and using (1.16) we find

$$\left(\sum_{i=1}^n c_i(v_{|F})^2 \right)^{1/2} \leq d_F \pi_2(v),$$

hence

$$c_i(v_{|F}) \leq d_F \pi_2(v) i^{-1/2}.$$

Clearly, we have

$$c_{i+j-1}(v) \leq c_i(v_{|F}).$$

Therefore, choosing $i = j = m$, we have, if $k = 2m - 1$,

$$c_k(v) \leq d_F \pi_2(v) 2^{1/2} m^{-1/2}$$
$$\leq D\left(\frac{n}{k} \right)^\alpha 2^{\alpha + 1/2} \pi_2(v) k^{-1/2}$$

and this also holds if $k = 2m$. ∎

Proof of Theorem 10.4: We prove (iv) $\Rightarrow$ (v). Assume (iv). Then every f.d. subspace $E \subset X$ satisfies the assumption of Lemma 10.5 for $\alpha > 1$ with a uniform constant D.

Now consider $v : E \to \ell_2^n$. By Lemma 10.5 we have

$$\sup_{k \leq n} k^{\alpha + 1/2} c_k(v) \leq D(2n)^{\alpha + 1/2} (\pi_2(v) n^{-1/2}).$$

By Carl's Theorem (cf. Theorem 5.2 above) this implies

$$\sup_{k \leq n} k^{\alpha + 1/2} e_k(v) \leq D' n^{\alpha + 1/2} (\pi_2(v) n^{-1/2})$$

for some constant D'. In particular, $e_n(v) \leq D' \pi_2(v) n^{-1/2}$.

This proves that (iv) $\Rightarrow$ (v). Let us prove that (v) $\Rightarrow$ (vi). Assume (v). Let $E \subset X$ be an arbitrary n-dimensional subspace. By Proposition 3.8 there is an operator $u : \ell_2^n \to E$ such that $\|u\| = 1$ and $\pi_2(u^{-1}) = n^{1/2}$ (in fact, $u(B_{\ell_2^n})$ is John's ellipsoid).

By (v) we have $e_n(u^{-1}) \leq C$, hence,

$$\frac{\mathrm{vol}(u^{-1}(B_E))}{\mathrm{vol}(B_{\ell_2^n})} \leq 2^n C^n;$$

equivalently, we have

$$vr(E) \leq \left(\frac{\mathrm{vol}(B_E)}{\mathrm{vol}(u(B_{\ell_2^n}))}\right)^{1/n} \leq 2C.$$

This shows that (v) $\Rightarrow$ (vi).

Finally, we use Theorem 10.2 and note that the implication (vi) $\Rightarrow$ (iii) follows immediately from Corollary 6.2.

Remark 10.6: We note in passing that X is a weak cotype 2 space iff X satisfies the following property:

(vii) There is a constant C_1 and $0 < \delta_1 < 1$ such that for all n, all n-dimensional subspaces $E \subset X$ and all $u : E \to \ell_2^n$ we have

$$c_{[\delta_1 n]}(u) \leq C_1 n^{-1/2} \pi_2(u).$$

Indeed, the implication (iv) $\Rightarrow$ (vii) follows from Lemma 10.5. Conversely, (vii) $\Rightarrow$ (ii) is easy. Indeed, if $E \subset X$ is n-dimensional, consider $u : \ell_2^n \to E$ such that $\|u\| \leq 1$ and $\pi_2(u^{-1}) \leq n^{1/2}$. Then if (vii) holds, there is a subspace $F \subset E$ with

$$\dim F \geq n - [\delta_1 n] \text{ such that } \|u_{|F}^{-1}\| \leq C_1.$$

This implies $d_F \leq C_1$ and (ii) follows. ∎

We now discuss the relationship between weak cotype 2 and the more usual notion of cotype q.

Proposition 10.7.

(1) *There is a numerical constant β such that for all n and all n-dimensional normed spaces E we have*

(10.11) $C_2(E) \leq \beta(1 + \mathrm{Log}\, n) wC_2(E).$

(2) *Let X be a weak cotype 2 space. Then X is of cotype q for every $q > 2$ and we have*

$$C_q(X) \le \beta(q)wC_2(X),$$

where $\beta(q)$ is a constant depending only on q.

Proof: We claim that there is a constant β such that for all m and all $u : \ell_2^m \to E$ we have

$$(10.12) \qquad \pi_2(u) \le \beta(1 + \text{Log } n) \sup k^{1/2} a_k(u).$$

This clearly implies (10.11) since we have, denoting $x_i = u(e_i)$,

$$\Big(\sum_1^m \|x_i\|^2 \Big)^{1/2} \le \beta(1 + \text{Log } n) wC_2(E)\ell(u).$$

To prove (10.12), we assume $\sup k^{1/2} a_k(u) = 1$. Let $v_k : \ell_2^m \to E$ be operators satisfying $rk(v_k) < 2^k$ and $\|u - v_k\| \le a_{2^k}(u)$. Let $\Delta_o = v_o$ and $\Delta_k = v_k - v_{k-1}$ so that $rk(\Delta_k) < 2^k + 2^{k+1}$ and $\|\Delta_k\| \le \|v_k - u\| + \|u - v_{k-1}\| \le 2^{-k/2} + 2^{-(k-1)/2} \le 4.2^{-k/2}$.

Note that $v_k = u$ if $2^k \ge n$ since $\dim E = n$, so that $u = \sum_{0 \le k < 1 + \text{Log}_2 n} \Delta_k$. This implies, by the triangle inequality (and the second remark following Proposition 3.8),

$$\begin{aligned}
\pi_2(u) &\le (1 + \log_2 n) \sup \pi_2(\Delta_k) \\
&\le (1 + \log_2 n) \sup_k (\|\Delta_k\|(rk(\Delta_k))^{1/2}) \\
&\le \beta(1 + \log n) \text{ for some numerical constant } \beta.
\end{aligned}$$

This proves (1). Let us now prove (2).

Let $C = wC_2(X)$. Let $x_1, \dots, x_m$ be an arbitrary sequence in X. Assume that

$$\Big\| \sum_1^m g_i x_i \Big\|_{L_2(X)} = 1.$$

We will prove that

$$(10.13) \qquad \Big(\sum_1^m \|x_n\|^q \Big)^{1/q} \le \beta(q)C.$$

We may as well assume $\|x_1\| \ge \|x_2\| \ge \cdots \ge \|x_m\|$.

Let $n \leq m$ be arbitrary, let E be the span of $x_1, \ldots, x_n$. By (10.11) (since $wC_2(E) \leq wC_2(X)$), we have

$$\left(\sum_1^n \|x_i\|^2\right)^{1/2} \leq \beta(1 + \text{Log } n)C\left\|\sum_1^n g_i x_i\right\|_{L_2(X)}$$

$$\leq \beta(1 + \text{Log } n)C.$$

Hence

$$\|x_n\| \leq n^{-1/2}\left(\sum_1^n \|x_i\|^2\right)^{1/2} \leq \beta C(1 + \text{Log } n)n^{-1/2},$$

and this implies (10.13), with

$$\beta(q) = \beta\left(\sum_{n=1}^{\infty}\left((1 + \text{Log } n)n^{-1/2}\right)^q\right)^{1/q}. \quad \blacksquare$$

Let $\alpha = (\alpha_i)$ be either a finite sequence of real numbers or an infinite sequence tending to zero at infinity. We will denote by (α_i^*) the non-increasing rearrangement of the sequence $(|\alpha_i|)_i$. Then we introduce

$$\|\alpha\|_{2\infty} = \sup_i i^{1/2}\alpha_i^*.$$

Equivalently, we have

$$\|\alpha\|_{2\infty} = \sup_{t>0} t(\text{card}\{i|\ |\alpha_i| > t\})^{1/2}.$$

We denote by $\ell_{2\infty}$ the space of all infinite sequences $\alpha = (\alpha_i)$ such that $\|\alpha\|_{2\infty} < \infty$.

Proposition 10.8. *Let X be a weak cotype 2 space. Let $x_1, x_2, \ldots, x_n$ be elements of X such that*

$$(10.14) \qquad \forall\ (\alpha_i) \in \mathbb{R}^n \ \sup|\alpha_i| \leq \left\|\sum_1^n \alpha_i x_i\right\|.$$

Then we have

$$(10.15) \qquad \sqrt{n} \leq 2wC_2(X)\left\|\sum_1^n g_i x_i\right\|_{L_2(X)},$$

and for all $\alpha = (\alpha_i)_{i \leq n}$ in $\mathbb{R}^n$

$$(10.16) \qquad \|\alpha\|_{2\infty} \leq 2wC_2(X)\left\|\sum_1^n \alpha_i g_i x_i\right\|_{L_2(X)}.$$

Proof: Let E be the subspace of X spanned by $x_1, \ldots, x_n$ and let $u : \ell_2^n \to X$ be the operator defined by $u(\alpha) = \sum \alpha_i e_i$ for all α in $\mathbf{R}^n$. We claim that $\pi_2(u^{-1}) \le n^{1/2}$. Indeed, this is easy since u^{-1} factors as $u^{-1} = jA$,

where j is the natural inclusion map and $\|A\| \le 1$ by (10.14). Since obviously $\pi_2(j) \le n^{1/2}$, we deduce $\pi_2(u^{-1}) \le n^{1/2}$.

We now use the weak cotype 2 assumption.

Let $C = wC_2(X)$. For every $k = 1, \ldots, n$, there is a subspace $S \subset \ell_2^n$ with dim $S > n - k$ such that

$$\|uP_S\| \le Ck^{-1/2}\ell(u).$$

(Here again P_S denotes the orthogonal projection of ℓ_2^n onto S.) Therefore we have

$$(\dim S)^{1/2} = \pi_2(u^{-1}uP_S) \le \pi_2(u^{-1})\|uP_S\|$$
$$\le n^{1/2}C\,k^{-1/2}\ell(u),$$

so that $(n - k + 1)^{1/2}k^{1/2}n^{-1/2} \le C\,\ell(u)$. Choosing $k = n/2$ if n is even and $k = (n+1)/2$ if n is odd, we obtain (10.15).

To prove (10.16) we may as well assume $|\alpha_1| \ge |\alpha_2| \ge \ldots \ge |\alpha_n|$. We note that the symmetry of the independent variables (g_i) implies that the sequence $(g_i x_i)$ is 1-unconditional in $L_2(X)$. Therefore we can write, for any k,

$$\left\|\sum_1^n \alpha_i g_i x_i\right\|_{L_2(X)} = \left\|\sum_1^n |\alpha_i| g_i x_i\right\|_{L_2(X)}$$
$$\ge \left\|\sum_1^k |\alpha_i| g_i x_i\right\|_{L_2(X)}$$
$$\ge |\alpha_k|\left\|\sum_1^k g_i x_i\right\|_{L_2(X)};$$

hence by (10.15)

$$\le |\alpha_k|(2C)^{-1}k^{1/2}.$$

Taking the supremum over k, we obtain (10.16). ∎

Let $\lambda \geq 1$. Recall that a finite or infinite sequence (x_i) in a Banach space is called λ-*unconditional* if, for all finitely supported sequences of scalars (α_i) and for all choices of signs $\xi_i = \pm 1$, we have

$$(10.17) \qquad \left\| \sum \xi_i \alpha_i x_i \right\| \leq \lambda \left\| \sum \alpha_i x_i \right\|.$$

The smallest λ for which this holds is called the *unconditionality constant* of $\{x_i\}$.

Corollary 10.9. *Let $x_1, \ldots, x_n$ be a normalized λ-unconditional sequence in a weak cotype 2 space X. Then we have, for all α in $\mathbf{R}^n$,*

$$(10.18) \qquad \|\alpha\|_{2\infty} \leq \lambda^2 C \left\| \sum_1^n \alpha_i x_i \right\|,$$

where C is a constant depending only on $wC_2(X)$ (and hence independent of n). The same holds for an infinite sequence $\{x_i | i \in \mathbf{N}\}$.

Remark: Let (ε_n) be a sequence of independent random variables taking the values $+1$ and -1 with probability $1/2$. We will use the following known fact: if X is of cotype $q < \infty$, there is a constant β depending only on q and on $C_q(X)$ such that for all n and all $y_1, \ldots, y_n$ in X we have

$$(10.19)$$
$$(2/\pi)^{1/2} \left\| \sum \varepsilon_i y_i \right\|_{L_2(Y)} \leq \left\| \sum g_i y_i \right\|_{L_2(Y)} \leq \beta \left\| \sum \varepsilon_i y_i \right\|_{L_2(Y)}.$$

The left-hand side is valid for any space X. We refer to [MaP] (cf. also [P1], p. 187) for a proof of this fact.

Proof of Corollary 10.9: Note that (since $\|x_i\| = 1$ by assumption) we have

$$\sup |\alpha_i| \leq \lambda \left\| \sum \alpha_i x_i \right\|.$$

Hence, by Proposition 10.8,

$$\|\alpha\|_{2\infty} \leq 2\lambda w C_2(X) \left\| \sum g_i \alpha_i x_i \right\|_{L_2(X)};$$

hence by (10.19),

$$\leq 2\lambda \beta w C_2(X) \left\| \sum \varepsilon_i \alpha_i x_i \right\|_{L_2(X)},$$

and by (10.17),

$$\leq 2\lambda^2 \beta w C_2(X) \left\| \sum \alpha_i x_i \right\|.$$

This proves (10.18). The second asertion of Corollary 10.9 is then immediate. ∎

Remark 10.10: It is worthwhile to note that when (x_i) is λ-unconditional, every normalized sequence of blocks of (x_n) is again λ-unconditional, and hence also satisfies (10.18) with the same constant C. Consider a sequence (ε_n) as above. We assume (ε_n) defined on some probability space $(\Omega, A, \mathbf{P})$. (For instance, the Lebesgue interval.) Let X be a Banach space; we will denote by $\mathrm{Rad}(X)$ the closed subspace of $L_2(\Omega, \mathbf{P}; X)$ spanned by all the X-valued functions of the form $\sum_{i=1}^{n} \varepsilon_i x_i (n \in \mathbf{N}, x_i \in X)$. Equivalently, $\mathrm{Rad}(X)$ is the space of all series $\sum_{n=1}^{\infty} \varepsilon_n x_n$ with coefficients in X which converge in $L_2(X) = L_2(\Omega, \mathbf{P}; X)$.

Similarly, we consider an i.i.d. sequence of Gaussian variables (g_n) on $(\Omega, A, \mathbf{P})$ and we denote by $G(X)$ the closure in $L_2(X)$ of all the functions $\sum_{i=1}^{n} g_i x_i$. Again, $G(X)$ coincides with the set of all the series $\sum_{n=1}^{\infty} g_n x_n$ which converge in $L_2(X)$. Note that with the notation introduced at the end of Chapter 2, $G(X)$ is the closure in $L_2(X)$ of $\cup_n G_n(X)$.

In the next statement we denote simply L_2 and $L_2(X)$ instead of $L_2(\Omega, \mathbf{P})$ and $L_2(\Omega, \mathbf{P}; X)$.

Proposition 10.11. *The following properties of a Banach space X are equivalent.*

(a) X *is a cotype 2 space.*
(b) $\ell_2(X)$ *is a weak cotype 2 space.*
(c) $L_2(X)$ *is a weak cotype 2 space.*
(d) $G(X)$ *is a weak cotype 2 space.*
(e) $\mathrm{Rad}(X)$ *is a weak cotype 2 space.*

Proof: It is easy to show that X is a cotype 2 space iff the same is true for $L_2(X)$ or $\ell_2(X)$. Hence (a) $\Rightarrow$ (b) is easy. The implication (b) $\Rightarrow$ (c) follows from the elementary fact that $L_2(X)$ is finitely representable into $\ell_2(X)$ (see the comments following Definition 10.1). The implication (c) $\Rightarrow$ (d) is trivial. The implication (d) $\Rightarrow$ (e) follows from the fact that if (d) holds, then $\mathrm{Rad}(X)$ and $G(X)$ are isomorphic. Indeed (d) implies X weak cotype 2, hence (cf. Proposition 10.7) of cotype q for $q > 2$, and this implies by (10.19) that $G(X)$ and $\mathrm{Rad}(X)$ are isomophic in a natural way.

Thus it remains to show (e) $\Rightarrow$ (a). Assume (e). We will use
Proposition 10.8. Consider $x_1, \ldots, x_n$ in X with $\|x_i\| = 1$. Then, for
any (α_i) in $\mathbf{R}^n$ we have clearly

$$\sup |\alpha_i| \leq \left\|\sum \alpha_i \varepsilon_i x_i\right\|_{L_2(X)}.$$

Let $C = 2wC_2(\mathrm{Rad}(X))$ and let us denote simply by $\| \ \|_2$ the norm in
$L_2(X)$. By (10.15) we have

$$n^{1/2} \leq C\left\|\sum_1^n g_i \varepsilon_i x_i\right\|_{L_2(\mathrm{Rad}(X))};$$

hence by symmetry,

(10.20) $$n^{1/2} \leq C\left\|\sum g_i x_i\right\|_2.$$

We now show that this implies, for all (α_i) in $\mathbf{R}^n$,

(10.21) $$\left(\sum |\alpha_i|^2\right) \leq C\left\|\sum_1^n g_i \alpha_i x_i\right\|_2.$$

Indeed, it is clearly enough to show this for α_i of the form $\alpha_i = (k_i/N)^{1/2}$ with k_i integers such that $\sum_{i=1}^n k_i = N$. To check this special case, let $(\overline{x}_i)_{i \leq N}$ be the sequence obtained from $(x_1, \ldots, x_n)$ after repeating x_1 k_1-times, repeating x_2 k_2-times, etc. Applying (10.20)
above, we find

(10.22) $$N^{1/2} \leq C\left\|\sum_1^N g_i \overline{x}_i\right\|_2,$$

but since $g_1 + \cdots + g_{k_1}$ coincides in distribution with $(k_1)^{1/2}g_1$ and
similarly for $g_{k_1+1} + \cdots + g_{k_1+k_2}$ and the other terms, we have

(10.23) $$\left\|\sum_1^N g_i \overline{x}_i\right\|_2 = \left\|\sum_1^n (k_i)^{1/2} g_i x_i\right\|_2.$$

Combining (10.22) and (10.23), we obtain (10.21) for $\alpha_i = (k_i/N)^{1/2}$.
Finally, (10.21) clearly implies that X is a cotype 2 space with
$C_2(X) \leq C$, and this concludes the proof that (e) implies (a). ∎

It is natural to ask whether one can develop similar ideas as in Theorem 10.2, but replacing $d_X(\delta)$ by a different quantity measuring the dimensions of (for instance) subspaces with unconditional basis (instead of Euclidean subspaces). The aim of the end of this chapter is to show that (rather surprisingly) this does *not* lead to a more general notion. We follow here the ideas of the paper [FJ1].

We first recall some terminology. We say that a Banach space X has the $g\ell_2$ property if there is a constant C such that for every 1-summing operator $u : X \to \ell_2$, there is an L_1-space $L_1(\mu)$ and. operators $A : X \to L_1(\mu)$ and $B : L_1(\mu) \to Y$ such that $u = BA$ and $\|A\| \, \|B\| \leq C\pi_1(u)$. The smallest constant C for which this holds will be denoted by $g\ell_2(X)$. It is known that every space with a λ-unconditional basis has the $g\ell_2$ property and $g\ell_2(X) \leq \lambda$.

More generally, every complemented subspace of a Banach lattice has the $g\ell_2$ property (cf. [GL]). The $g\ell_2$ property is clearly stable under isomorphisms.

The following lemma from [FJ1] is crucial for our next result.

Lemma 10.12.   *There are absolute constants λ and α with $0 < \alpha < 1$ such that for every n, every n-dimensional normed space E and every $u : E \to \ell_2^n$, there is a subspace $F \subset E$ with $\dim F \geq \alpha n$ such that $\pi_1(u_{|F}) \leq \lambda \pi_2(u)$.*

We refer to [FJ1], p. 98, for the proof.

The next result follows from the results of [FJ1]; it was communicated to the author by W. B. Johnson.

Theorem 10.13.   *The following properties of a Banach space X are equivalent.*

(i) X *is a weak cotype 2 space.*

(viii) *There is a constant C and $0 < \delta < 1$ such that for all n every n-dimensional subspace $E \subset X$ contains a subspace $F \subset E$ with $\dim F \geq \delta n$ such that $g\ell_2(F) \leq C$.*

Proof: The implication (i) $\Rightarrow$ (viii) follows trivially from Theorem 10.2 since (ii) $\Rightarrow$ (viii) is immediate (indeed note that $d_F \leq C$ implies $g\ell_2(F) \leq C$). To prove the converse, assume (viii). We will show that (vii) holds (cf. Remark 10.6). Consider an n-dimensional subspace

$E \subset X$ and $u : E \to \ell_2^n$. By the preceding lemma, $\exists\, F \subset E$ with $\dim F \geq \alpha n$ such that

(10.24) $\pi_1(u_{|F}) \leq \lambda \pi_2(u).$

By (viii) there is a further subspace $F_1 \subset F$ with $\dim F_1 \geq \alpha \delta n$ such that $u_{|F_1}$ admits a factorization $u_{|F_1} = BA$ with $A : F_1 \to L_1(\mu)$ and $B : L_1(\mu) \to \ell_2^n$, satisfying

(10.25) $\|B\|\, \|A\| \leq C\pi_1(u_{|F_1}) \leq C\pi_1(u_{|F}).$

(Consult the diagram below)

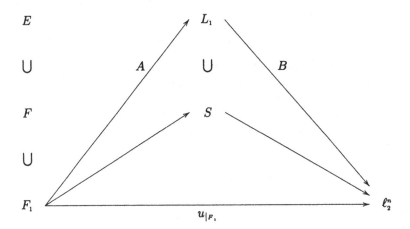

We may assume without loss of generality that u is an isomorphism. Denote $m = \dim F_1$. We then introduce a subspace $S \subset L_1(\mu)$ such that $S = A(F_1)$ ($\dim S = m$), and we apply Remark 10.6 this time to the Banach space $L_1(\mu)$ which is a cotype 2 (hence *a fortiori* weak cotype 2) space. We thus obtain from Remark 10.6 constants C_1 and $0 < \delta_1 < 1$ such that

$$c_{[\delta_1 m]}(B_{|S}) \leq C_1 m^{-1/2} \pi_2(B_{|S})$$
$$\leq C_1 m^{-1/2} \pi_2(B).$$

By (a weak form of) Grothendieck's Theorem, it is known (cf. e.g. [P2], p. 57) that $\pi_2(B) \leq (\pi/2)^{1/2}\|B\|$ for all $B : L_1(\mu) \to \ell_2^n$. Thus we obtain finally

$$c_{[\delta_1 m]}(B_{|S}) \leq C_1(\pi/2)^{1/2}m^{-1/2}\|B\|,$$

which implies, since $u_{|F_1} = B_{|S}A_{|F_1}$,

$$c_{[\delta_1 m]}(u_{|F_1}) \leq C_1(\pi/2)^{1/2}\|A\|\,\|B\|m^{-1/2};$$

therefore, by (10.25) and (10.24),

$$c_{[\delta_1 m]}(u_{|F_1}) \leq C_1 C\lambda(\pi/2)^{1/2}\pi_2(u)m^{-1/2}.$$

This implies $\exists\, F_2 \subset F_1$ with $\dim F_2 \geq m - \delta_1 m$ such that $\|u_{|F_2}\| \leq C_3\pi_2(u)m^{-1/2}$ for some constant C_3. Finally, recalling that $m \geq \alpha\delta n$ we conclude that X satisfies (vii) and this completes the proof. ∎

Remark: The reader will easily check that we can replace $L_1(\mu)$ in the preceding argument by any weak cotype 2 space L such that every operator $B : L \to \ell_2$ is 2-summing. We can thus relax the $g\ell_2$-property by replacing $L^1(\mu)$ by L, and the preceding statement remains valid.

Remark: Finally, let us briefly return to the case of cotype q spaces. Let X be a Banach space of cotype q ($2 < q < \infty$). Then, as seen at the beginning of this chapter, there is a constant C such that for all n and all $u : \ell_2^n \to X$ we have

$$(10.26) \qquad\qquad \sup k^{1/q}a_k(u) \leq C\ell(u).$$

Now let $E \subset X$ be a subspace of dimension n. Let $u : \ell_2^n \to E$ be such that $\ell(u) = \ell^*(u^{-1} = n^{1/2})$ (cf. Theorem 3.1). Let P be any orthogonal projection on ℓ_2^n; we have, clearly,

$$rk(P) = tr\ P \leq \ell(uP)\ell^*(u^{-1}),$$

hence

$$\ell(uP) \geq rk(P)n^{-1/2}$$

for some numerical constant $\alpha > 0$.

In other words, we have found an E-valued Gaussian variable Z with $d(Z) \geq \alpha'n^{2/q}$ for some constant $\alpha' > 0$. Equivalently, $\delta(E) \geq \alpha'n^{2/q}$ (with the notation of Chapter 4). Therefore, by Theorem 4.4, we can state

Theorem 10.14.   *Let X be a space of cotype q (or merely satisfying (10.26) above). Then there is a function $\eta(\varepsilon) > 0$ such that every $E \subset X$ of dimension n contains a subspace $F \subset E$ with $\dim F = [\eta(\varepsilon)n^{2/q}]$ which is $(1 + \varepsilon)$-isomorphic to a Euclidean space.*

The preceding result shows that if a space X is roughly *far from* L_∞ then the Log N estimate in Dvoretzky's Theorem (cf. Chapter 4) can be improved to a power of N.

We recall that for example L_q is of cotype q if $q \geq 2$ (and of cotype 2, otherwise) so that Theorem 10.14 applies in that case. This refines the estimate given above in Proposition 4.15.

Remark: It seems natural to call the spaces satisfying (10.26) *weak cotype q spaces*. But actually, the class of spaces which are really of interest to us from the point of view of these notes are the spaces which satisfy the conclusion of Theorem 10.14, and for these we know of no neat characterization so far.

Notes and Remarks

In their paper on the inverse Santaló inequality (cf. Chapter 7), Bourgain and Milman proved that any space of cotype 2 has a uniformly bounded volume ratio. In other words, they proved that every space of cotype 2 satisfies the property (vi) in the preceding chapter.

This result was the main motivation for the study of weak cotype 2 spaces which was developed in [MP1].

Thus all the statements from Definition 10.1 to Proposition 10.11 come from [MP1], with minor refinement occasionally. Note, however, that we have incorporated the estimate of [PT1] to state the best known estimate for $d_X(\delta)$ in Theorem 10.2.

Theorem 10.13 is due to Figiel and Johnson; it is implicit in [FJ1] and was (explicitly) communicated to us by W. B. Johnson. Finally, Theorem 10.14 is due to Figiel, Lindenstrauss, and Milman [FLM].

Concerning the notion of weak cotype q, it was observed by U. Matter and V. Mascioni that this is equivalent to the cotype q property restricted to vectors of equal norm (precisely: there is a constant C such that for any $x_1, \dots, x_n$ in the unit sphere of X we have $n^{1/q} \leq C\| \sum g_k x_k \|_{L_2(X)}$). It is known that the latter property is strictly weaker than cotype q if $q > 2$. This follows from an example of Tzafriri; see [CS] for more details.

Chapter 11

Weak Type 2

This chapter is based on [MP1] and [Pa1]. We proceed as in the previous chapter and start by recalling the notion of type p. Let $1 \leq p \leq 2$. A Banach space X is called of type p if there is a constant C such that for all n and all $x_1, \ldots, x_n$ in X we have

$$(11.1) \qquad \left\| \sum g_i x_i \right\|_{L_2(X)} \leq C \left(\sum \|x_i\|^p \right)^{1/p}.$$

The smallest constant C for which this holds is called the type p constant of X and is denoted by $T_p(X)$. As for cotype q, the above definition is equivalent to the same notion with a sequence (ε_n) of independent symmetric r.v.'s with values in $\{-1, +1\}$. Again we refer to [MaP], [P1], and [MS] for more information.

Let $u : \ell_2^n \to X$ be an arbitrary operator. Let (f_i) be an arbitrary orthonormal basis of ℓ_2^n. Clearly, (11.1) can be rewritten as

$$\ell(u) \leq C \left(\sum \|u f_i\|^p \right)^{1/p}.$$

By a simple duality argument, this is equivalent to the following:

$$\forall \, v : X \to \ell_2^n \qquad \left(\sum \|v^* f_i\|^q \right)^{1/q} \leq C \, \ell^*(v),$$

with $1/q + 1/p = 1$.

By Lemma 1.8, this inequality implies

$$\left(\sum a_i(v)^q \right)^{1/q} = \left(\sum a_i(v^*)^q \right)^{1/q} \leq C \, \ell^*(v),$$

and, *a fortiori*,

$$\sup_{k \geq 1} k^{1/q} a_k(v) \leq C \, \ell^*(v).$$

169

We thus come to the following:

Definition 11.1. *Let X be a Banach space. We say that X is a weak type 2 space (or briefly X is weak type 2) if there is a constant C such that, for all n and all operators $v : X \rightarrow \ell_2^n$ we have*

$$(11.2) \qquad\qquad \sup_k k^{1/2} a_k(v) \leq C \, \ell^*(v).$$

The smallest constant C for which this holds will be denoted by $wT_2(X)$.

By the above remarks, type 2 $\Rightarrow$ weak type 2 and $wT_2(X) \leq T_2(X)$. It is rather easy to see that weak type 2 is inherited by the subspaces and the quotient spaces of a weak type 2 space. More generally, any space which is finitely representable in a weak type 2 space is again weak type 2. Also, the direct sum of a *finite* number of weak type 2 spaces is again one.

The following two propositions will allow us to reduce several questions concerning weak type 2 to the analogous questions for weak cotype 2.

We start by a simple statement.

Proposition 11.2.

(1) *There is a numerical constant β such that for all n and all n-dimensional spaces E we have*

$$T_2(E) \leq \beta(1 + \text{Log n})wT_2(E).$$

(2) *Let X be a weak type 2 space. Then X is type p for every $p < 2$ and we have*
$$T_p(X) \leq \beta(p)wT_2(X)$$

for some constant $\beta(p)$ depending only on p.

Proof: This can be proved exactly as Proposition 10.7.

We will use several times in this chapter the following result from [P3] (already mentioned in Chapter 2) which we state without proof. (See also [P1] or [MS].)

Theorem 11.3. *A Banach space X is not K-convex iff for each $\varepsilon > 0$ and each $n \geq 1$ there is a subspace $E \subset X$ such that $d(E, \ell_1^n) < 1 + \varepsilon$.*

(In that case, we say that X contains ℓ_1^n's uniformly.) Moreover, X is K-convex iff X is of type p for some $p > 1$. Also, X is K-convex iff X is locally π-Euclidean, which means the following: There is a constant C and for each $\varepsilon > 0$ and $n \geq 1$ an integer $N(n, \varepsilon)$ such that every subspace $E \subset X$ with dim $E \geq N(n, \varepsilon)$ contains a subspace $F \subset E$ with dim $F = n$ such that $d_F \leq 1 + \varepsilon$ and F is C-complemented in X (which means there is a projection $Q : X \to F$ with $\|Q\| \leq C$).

Proposition 11.4. *Let X be a Banach space. Then X is weak type 2 iff X^* is K-convex and weak cotype 2. Moreover, $wC_2(X^*) \leq wT_2(X) \leq K(X)wC_2(X^*)$. On the other hand, X^* is weak type 2 iff X is K-convex and weak cotype 2. (We recall that K-convexity is a self-dual property and $K(X) = K(X^*)$.)*

Proof: We have seen in Chapter 3 that if X is K-convex, then for all $v : X \to \ell_2^n$ we have

$$(11.3) \qquad \ell(v^*) \leq K(X)\ell^*(v).$$

Thus, if X^* is K-convex and weak cotype 2, we have

$$\sup k^{1/2} a_k(v^*) \leq wC_2(X^*)\ell(v^*)$$
$$\leq wC_2(X^*)K(X)\ell^*(v).$$

This shows that X is weak type 2 and $wT_2(X) \leq wC_2(X^*)K(X)$. Conversely, assume X weak type 2. Then, by Proposition 11.2, X is of type $p > 1$, hence by Theorem 11.3 X is K-convex. Now using the trivial inequality $\ell^*(v) \leq \ell(v^*)$ for all $v : X \to \ell_2^n$, we deduce from (11.2) that X^* is weak cotype 2 and $wC_2(X^*) \leq wT_2(X)$. This proves the first part of the statement. It is easy to see that we can interchange the roles of X and X^* in the above proof, whence the second part. ∎

Remark 11.5: We have just proved that X is weak type 2 iff it is K-convex and $d_{X^*}(\delta) < \infty$ for all $0 < \delta < 1$. We should emphasize that if X is K-convex, we may drop the logarithmic term in the upper estimate of $d_{X^*}(\delta)$ in Theorem 10.2 above, so that we have $d_{X^*}(\delta) \leq K(1 - \delta)^{-1}$ for some constant K. This is clear from the proof of Theorem 10.2 (recall that the *log* factor in Theorem 10.2 comes from the estimate of finite-dimensional K-convexity constants which are unnecessary if X and X^* are K-convex.)

While weak cotype 2 is related (via Theorem 10.2) to the existence of *large* Euclidean sections, the notion of weak type 2 is linked to the existence of projections of large ranks onto Euclidean subspaces. This direction was motivated by a result of Maurey [Ma1], who proved that if X is type 2 then every operator $u : S \to \ell_2$ defined on a subspace S of X admits an extension $\widetilde{u} : X \to \ell_2$. Equivalently, there is a constant C such that for all $S \subset X$ and all $u : S \to \ell_2^n$ there is an extension $\widetilde{u} : X \to \ell_2^n$ such that $\|\widetilde{u}\| \leq C\|u\|$.

The converse is still an open question. We will see, however, that a weak form of this extension property characterizes weak type 2. (In this chapter the properties numbered from (i) to (ix) are all equivalent to weak type 2.)

Theorem 11.6.   *The following properties of a Banach space X are equivalent.*

(i) *X is weak type 2.*

(ii) *There is a constant C satisfying for any subspace $S \subset X$ and any $u : S \to \ell_2^n$ there is an extension $\widetilde{u} : X \to \ell_2^n$ such that for any $k = 1, 2, \ldots, n$ there is a projection $P : \ell_2^n \to \ell_2^n$ of rank $> n - k$ such that $\|P\widetilde{u}\| \leq C(n/k)^{1/2}\|u\|$.*

(iii) *There is a $0 < \delta < 1$ and a constant C such that for any $S \subset X$ and any $u : S \to \ell_2^n$ there is a projection $P : \ell_2^n \to \ell_2^n$ of rank $> \delta n$ and an operator $v : X \to \ell_2^n$ with $v_{|S} = Pu$ such that $\|v\| \leq C\|u\|$.*

(iv) *There is a constant C and $0 < \delta < 1$ such that every f.d. quotient space Z of X satisfies the following: for every n, every n-dimensional subspace $E \subset Z$ with $d_E \leq 2$ contains a subspace $F \subset E$ with $\dim F \geq \delta n$ onto which there is a projection $Q : Z \to F$ with $\|Q\| \leq C$.*

For the proof, the following *lifting lemma* already used in Chapter 9 will be useful.

Lemma 11.7.   *Let S be a closed subspace of a K-convex space X. Let $\sigma : X \to X/S$ denote the quotient mapping. Then, for each $\varepsilon > 0$ and all n, every operator $v : \ell_2^n \to X/S$ admits a lifting $\widetilde{v} : \ell_2^n \to X$ such that $\sigma\widetilde{v} = v$ and*

$$\ell(\widetilde{v}) \leq K(X)(1 + \varepsilon)\ell(v).$$

Proof: This follows immediately from Lemma 9.4.

Proof of Theorem 11.6: We first prove (i) $\Rightarrow$ (ii). Assume (i). Consider $u : S \to \ell_2^n$ and its adjoint $u^* : \ell_2^n \to X^*/S^\perp$. Let $\sigma : X^* \to X^*/S^\perp$ be the quotient mapping. Note the trivial estimate $\ell(u^*) \leq n^{1/2}\|u\|$. By the preceding remark, there is an operator $\tilde{v} : \ell_2^n \to X^*$ such that $\sigma\tilde{v} = u^*$ and $\ell(\tilde{v}) \leq 2n^{1/2}\|u\|K(X)$. Let $\tilde{u} = (\tilde{v})^*_{|X}$. Then $\tilde{u} : X \to \ell_2^n$ satisfies $\tilde{u}_{|S} = u$ and (since $\tilde{u}^* = \tilde{v}$)

$$(11.4) \qquad \ell(\tilde{u}^*) \leq 2n^{1/2}\|u\|K(X).$$

Since X^* is weak cotype 2 (and $a_k(\tilde{u}) = a_k(\tilde{u}^*)$), we have

$$\sup k^{1/2}a_k(\tilde{u}) \leq wC_2(X^*)\ell(\tilde{u}^*);$$

hence there is a projection P on ℓ_2^n with rank $> n - k$ such that

$$\|P\tilde{u}\| \leq wC_2(X^*)\ell(\tilde{u}^*)k^{-1/2}.$$

Recalling (11.4), we obtain (ii).

(ii) $\Rightarrow$ (iii) is trivial.

(iii) $\Rightarrow$ (iv): Assume (iii) and let $E \subset Z$ be as in (iv). We have an isomorphism $T : E \to \ell_2^n$ such that $\|T\| = 1$ $\|T^{-1}\| \leq 2$. Let $\sigma : X \to Z$ denote the quotient mapping. Let $S = \sigma^{-1}(E)$ and let $u : S \to \ell_2^n$ be defined by $u = T\sigma$. By (iii) there is a projection P on ℓ_2^n with rank $> \delta n$ and $v : X \to \ell_2^n$ such that $v_{|S} = Pu$ and $\|v\| \leq C\|u\|$. Let $F \subset E$ be the range of $T^{-1}P$. We will show that F is $2C$-complemented in Z. Note that u (and hence v) vanishes on $\operatorname{Ker}\sigma$. Therefore, v induces canonically a mapping $\tilde{v} : Z \to \ell_2^n$ such that $v = \tilde{v}\sigma$ and $\|\tilde{v}\| = \|v\|$.

Since $v_{|S} = Pu$, we have $\tilde{v}_{|E} = PT$. Then it is easy to check that $Q = T^{-1}P\tilde{v}$ is a projection from Z onto F and

$$\|Q\| \leq \|T^{-1}\| \, \|\tilde{v}\| \leq 2C\|u\|.$$

This shows that (iii) $\Rightarrow$ (iv).

(iv) $\Rightarrow$ (i): Assume (iv). We first note that (iv) (together with Dvoretzky's Theorem) implies that X is locally π-Euclidean, whence by Theorem 11.3 X is K-convex.

We will now show that X^* is weak cotype 2. Let $E \subset X^*$ be an n-dimensional subspace. By the QS-Theorem and Dvoretzky's Theorem, there is a fixed $0 < \delta' < 1$ for which there are subspaces $E_2 \subset E_1 \subset E$ with dim $E_1/E_2 \geq \delta'n$ such that $d_{E_1/E_2} \leq 2$. Then E_1^*

can be identified with $Z = X/E_1^{\perp}$. Also, $(E_1/E_2)^*$ can be identified with the subspace G of Z formed by the elements which vanish on E_2.

By (iv) there is a subspace $G_1 \subset G$ with $\dim G_1 \geq \delta \dim G$ which is C-complemented in Z by a projection $Q : Z \to G_1$. This clearly implies that Z admits a quotient space, namely $Z_1 = Z/\operatorname{Ker} Q$, which is C-isomorphic to G_1 and hence satisfies $d_{Z_1} \leq 2C$. Dualizing again, this means that $Z^* = E_1$ admits a subspace F with $\dim F \geq \delta \dim G$ (hence $\dim F \geq \delta \delta' n$) such that $d_F \leq 2C$. We have thus shown that X^* satisfies (ii) in Theorem 10.2, hence X^* is weak cotype 2. By Proposition 11.4, this completes the proof that (iv) $\Rightarrow$ (i). ∎

Remark: Note that in (iv) $\Rightarrow$ (i) we only use (iv) for $E \subset Z$ with $\dim E$ *proportional* to $\dim Z$.

Remark 11.8: In analogy with Proposition 10.11, we note the equivalence of the following properties:

(a) X is type 2.
(b) $\ell_2(X)$ is weak type 2.
(c) $L_2(X)$ is weak type 2.
(d) $G(X)$ is weak type 2.
(e) $\operatorname{Rad}(X)$ is weak type 2.

Indeed, when X is K-convex, we may identify $G(X)^*$ (resp. $\operatorname{Rad}(X)^*$) with $G(X^*)$ (resp. $\operatorname{Rad}(X^*)$), so that this follows immediately from Proposition 11.4 and Proposition 10.11.

We will now dualize Corollary 10.9. For that we need to introduce a predual of $\ell_{2\infty}$, namely the space ℓ_{21}. Let $\alpha = (\alpha_n)$ be a sequence of numbers tending to zero at infinity. Let (α_n^*) be the non-increasing rearrangement of $(|\alpha_n|)_n$. We define

$$\|\alpha\|_{21} = \sum_{n=1}^{\infty} \alpha_n^* n^{-1/2}.$$

We denote by ℓ_{21} the space of all sequences (α_n) such that $\|\alpha\|_{21} < \infty$. It becomes a Banach space when equipped with this norm. It is well known that $\ell_{21}^* = \ell_{2\infty}$ with equivalent norms.

Proposition 11.9. *Let $x_1, \ldots, x_n$ be a normalized λ-unconditional sequence in a weak type 2 space X. Then we have for all α in $\mathbf{R}^n$*

$$\left\|\sum \alpha_i x_i\right\| \leq \lambda^2 C \|\alpha\|_{21},$$

where C is a constant depending only on $wT_2(X)$. Moreover, the same result is valid for an infinite λ-unconditional sequence $\{x_n | n \geq r\}$.

Proof: Let E be the span of $x_1, \dots, x_n$. Let $x_1^*, \dots, x_n^*$ be the functionals in E^* which are biorthogonal to $x_1, \dots, x_n$. We have, by Proposition 11.4, $wC_2(E^*) \leq wT_2(E)$, hence by Corollary 10.9

$$\forall \, \alpha \in \mathbf{R}^n \quad \|\alpha\|_{2\infty} \leq \lambda^2 C \left\| \sum \alpha_i x_i^* \right\|.$$

By duality, this yields

$$\left\| \sum \alpha_i x_i \right\| \leq \lambda^2 C \|\alpha\|_{21} \quad \blacksquare$$

Remark: Curiously, Proposition 10.8 does not seem to *dualize* in a nice way.

We now present (divided into several pieces) the dual form of Theorem 10.4.

Theorem 11.10. *The following properties of a Banach space X are equivalent:*

(i) *X is a weak type 2 space.*

(v) *There is a constant C such that for all n and all $v : X^* \to \ell_2^n$ we have for $k = 1, 2, \dots, n$*

$$c_k(v) \leq C(n/k)^2 (\pi_2(v) n^{-1/2}).$$

(v)' *There is a constant C such that for all n and all $v : X^* \to \ell_2^n$ we have*

$$\max\{e_n(v^*), e_n(v)\} \leq C \, \pi_2(u) n^{-1/2}.$$

Proof: Assume (i) and consider $v : X^* \to \ell_2^n$. We can find a quotient space Z of X^* with $\dim Z = n$ such that v factors as $v = \tilde{v}\sigma$ where $\sigma : X^* \to Z$ is the quotient mapping and $\|\tilde{v}\| = \|v\|$. Note that by Proposition 11.4 we have (since Z^* is a subspace of X)

$$wC_2(Z) \leq wT_2(Z^*)$$
$$\leq wT_2(X).$$

By Remark 11.5, there is a fixed constant C such that for all $k = 1, \ldots, n$ there is a subspace $F \subset Z$ with $\dim F > n - k$ and $d_F \leq C(n/k)$. Let $m = \dim F > n - k$. Let $w : \ell_2^m \to F \subset Z$ be an isomorphism such that

$$(11.5) \qquad \|w\| \, \|w^{-1}\| \leq C(n/k)$$

Then, by Theorem 11.6, there is an operator $\widetilde{w} : \ell_2^m \to X^*$ and a projection P on ℓ_2^m with $\mathrm{rk}(P) \geq m - k$ such that $\sigma \widetilde{w} P = wP$ and

$$(11.6) \qquad \|\widetilde{w}P\| \leq C'(m/k)^{1/2}\|w\|$$

for some constant C'.

Then $v\widetilde{w}P : \ell_2^m \to \ell_2^m$ satisfies (recall (1.13) and Proposition 1.6)

$$\left(\sum a_k(v\widetilde{w}P)^2\right)^{1/2} = \|v\widetilde{w}P\|_{HS} = \pi_2(v\widetilde{w}P) \leq \pi_2(v)\|\widetilde{w}P\|;$$

hence $a_k(v\widetilde{w}P) \leq \pi_2(v)\|\widetilde{w}P\|k^{-1/2}$. Since $v\widetilde{w}P = (\widetilde{v}\sigma)\widetilde{w}P = \widetilde{v}wP$ we obtain

$$a_k(\widetilde{v}wP) \leq \pi_2(v)\|\widetilde{w}P\|k^{-1/2}.$$

Let F_1 be the range of wP. We have $\widetilde{v}_{|F_1} = \widetilde{v}(wP)w_{|F_1}^{-1}$, hence

$$(11.7) \qquad c_k(\widetilde{v}_{|F_1}) \leq \|w^{-1}\|\pi_2(v)\|\widetilde{w}P\|k^{-1/2},$$

and

$$(11.8) \qquad c_k(v_{|\sigma^{-1}(F_1)}) \leq c_k(\widetilde{v}_{|F_1}).$$

Note that codim $\sigma^{-1}(F_1) < \dim F/F_1 + \dim Z/F < 2k$, therefore

$$c_{3k}(v) \leq c_k(v_{|\sigma^{-1}(F_1)}),$$

so that (11.8), (11.7), (11.6), and (11.5) imply

$$c_{3k}(v) \leq CC'\pi_2(v)(m/k)^{1/2}(n/k)k^{-1/2}$$
$$\leq C''(n/k)^2(\pi_2(v)n^{-1/2}),$$

which immediately implies (v). This completes the proof that (i) $\Rightarrow$ (v).

Clearly (v) $\Rightarrow$ (v)' follows from Carl's Theorem, i.e. Theorem 5.2 above.

Let us show that (v)' $\Rightarrow$ (i). First, (v)' implies that X^* is a weak cotype 2 space. This follows from Theorem 10.5 and the fact that every operator $T : E \to \ell_2^n$ with $E \subset X^*$ admits an extension $\widetilde{T} : X^* \to \ell_2^n$ satisfying $\pi_2(\widetilde{T}) = \pi_2(T)$. Moreover, (v)' also implies that X is K-convex. Indeed, assume not. Then, by Theorem 11.3, X contains ℓ_1^n's uniformly, hence we can find a sequence $\{Z_n\}$ of quotient spaces of X^* and isomorphisms $v_n : Z_n \to \ell_\infty^n$ with $\|v_n\| = 1, \|v_n^{-1}\| \leq 2$.

Let $\sigma_n : X^* \to Z_n$ denote the quotient map and let $i_n : \ell_\infty^n \to \ell_2^n$ be the identity map. Clearly $\pi_2(i_n) \leq \sqrt{n}$ (actually equality holds), hence $\pi_2(i_n v_n \sigma_n) \leq \|v_n\| n^{1/2} = n^{1/2}$. Assuming (v)', we find

$$(11.9) \qquad e_n(i_n v_n \sigma_n) \leq C.$$

On the other hand, there is a numerical constant K such that we have (we assume for simplicity that K is an integer)

$$(11.10) \qquad e_{Kn}(i_n^{-1}) \leq K \, n^{-1/2}.$$

This elementary fact follows from Lemma 7.5 and the fact that by (1.18)

$$n^{-1/2} B_{\ell_\infty^n} \subset B_{\ell_2^n} \text{ and } \mathrm{vol}(B_{\ell_2^n})^{1/n} \leq K' \, \mathrm{vol}(n^{-1/2} B_{\ell_\infty^n})^{1/n}$$

for some numerical constant K'.

We thus conclude that by (11.9) and (11.10)

$$e_{n+[Kn]}(v_n \sigma_n) \leq e_n(i_n v_n \sigma_n) e_{[Kn]}(i_n^{-1})$$
$$\leq C K' n^{-1/2}.$$

This immediately implies (by (7.5))

$$\left(\frac{\mathrm{vol}(\sigma_n(B_{X^*}))}{\mathrm{vol}(B_{Z_n})} \right)^{1/n} \leq K'' n^{-1/2},$$

for some numerical constant K'' and this is absurd since the left side is equal to 1. This contradiction shows that (v)' implies X is K-convex and this concludes the proof that (v)' $\Rightarrow$ (i) by Proposition 11.4. ∎

We now turn to the results of [Pa1] which give a volume ratio characterization of weak type 2 spaces. For this we will use a result from [Ca2].

Lemma 11.11. *For all operators* $u : X \rightarrow Y$ *between Banach spaces, we have for all* n

$$c_n(u) \leq (\Pi_{k=1}^n c_k(u))^{1/n} \leq sup\left\{\left|\det(< y_i^*, u x_j >)\right|^{1/n}\right\}$$

where the supremum runs over all n-*tuples* $x_1, \ldots, x_n$ *in* B_X *and* $y_1^*, \ldots, y_n^*$ *in* B_{Y^*}.

Proof: Let $\varepsilon > 0$. There is x_1 in B_X and y_1^* in B_{Y^*} such that

$$< y_1^*, u x_1 > \geq \|u\|(1 - \varepsilon) = c_1(u)(1 - \varepsilon).$$

Let $S_1 = (y_1^*)^{\perp}$. Since $\|u_{|S_1}\| \geq c_2(u)$ there are x_2 in $B_X \cap S_1$ and y_2^* in B_{Y^*} such that

$$< y_2^*, u x_2 > > c_2(u)(1 - \varepsilon).$$

Continuing in this way we find for each $k \leq n$ x_k in $B_X \cap (y_1^*, \ldots y_{k-1}^*)^{\perp}$ and y_k^* in B_{Y^*} such that

$$< y_k^*, u x_k > \geq c_k(u)(1 - \varepsilon).$$

Consider the $n \times n$ matrix

$$A = (< y_i^*, u x_j >)_{i,j \leq n}.$$

By construction, this is a triangular matrix, hence

$$|\det(A)| = \left|\Pi < y_k^*, u x_k >\right| \geq \prod_1^n c_k(u)(1 - \varepsilon)^n.$$

This completes the proof. ∎

We briefly return to weak cotype 2 spaces in the next result from [Pa1].

Theorem 11.12. *The following assertions are equivalent:*

(a) *X is a weak cotype 2 space.*
(b) *There is a constant C such that for all n, all n-dimensional subspaces $E \subset X$ and all operators $u : E \rightarrow \ell_\infty^n$ we have*

$$e_{[Cn]}(u) \leq Cn^{-1/2}\|u\|.$$

(c) *There is a constant C_1 such that for all n, all n-dimensional subspaces $E \subset X$, and all operators $u : E \to \ell_\infty^n$ we have*

$$(11.11) \qquad \left(\frac{\mathrm{vol}(u(B_E))}{\mathrm{vol}(B_{\ell_\infty^n})} \right)^{1/n} \le C_1 n^{-1/2} \|u\|.$$

Proof: The implication (a) $\Rightarrow$ (b) is a simple consequence of the results of Chapter 10. Indeed, assume (a). Let $i_n : \ell_\infty^n \to \ell_2^n$ be the identity map. We have clearly $\pi_2(i_n) \le n^{1/2}$. Hence, by Theorem 10.4 for some constant C' we have

$$e_n(i_n u) \le C' \pi_2(i_n u) n^{-1/2}$$
$$\le C' \pi_2(i_n) \|u\| n^{-1/2} \le C' \|u\|.$$

Now let us write $u = i_n^{-1}(i_n u)$ and recall that by (11.10) above we have

$$e_{[Kn]}(i_n^{-1}) \le K n^{-1/2}$$

for some absolute constant K. This implies

$$e_{n+[Kn]}(u) \le e_{[Kn]}(i_n^{-1}) e_n(i_n u)$$
$$\le K \, n^{-1/2} C' \|u\|.$$

Choosing $C = \max(KC', 1 + K)$ we obtain (b). This shows that (a) $\Rightarrow$ (b). The implication (b) $\Rightarrow$ (c) is obvious (using (7.5)). Let us prove (c) $\Rightarrow$ (a). Assume (11.11). We will show that (10.1) holds. Consider $u : \ell_2^n \to X$. We may assume for simplicity that u is an isomorphism of ℓ_2^n onto $E \subset X$ with $\dim E = n$. We will show that (11.11) implies for some constant C for all $k \ge 1$ for all $(y_i^*)_{i \le k}$ in $B_{\ell_2^n}$ and all $(x_j)_{j \le k}$ in B_E,

$$(11.12) \qquad \left| \det(< y_i^*, u x_j >) \right|^{1/k} \le C \, k^{-1/2} \ell(u).$$

By Lemma 11.11 (since here $c_k(u) = a_k(u)$) this implies (10.1) and X is a weak cotype 2 space.

It remains to check (11.12). Let us denote by H a k-dimensional subspace containing $x_1, \ldots, x_k$ and let $F = u(H)$. We know by Corollary 5.12 that

$$(11.13) \qquad e_k(u_{|H}) \le c_2 k^{-1/2} \ell(u_{|H}) \le c_2 \, k^{-1/2} \ell(u)$$

for some constant c_2.

We denote by $A : \ell_2^k \to H$ and $B : F \to \ell_\infty^k$ the operators defined by $Ae_i = x_i$ and $B(x) = (y_i^*(x))$. Clearly, we have by (1.13)

$$\pi_2(A) = \|A\|_{HS} = \left(\sum \|x_i\|^2\right)^{1/2} \leq k^{1/2};$$

hence, by Theorem 5.2,

(11.14) $e_k(A) \leq c_3$

for some constant c_3. Moreover, we have $\|B\| = \sup \|y_i^*\| \leq 1$. Now consider again $i_k : \ell_\infty^k \to \ell_2^k$ and the composition $\widetilde{A} = Ai_k$. Since $\|i_k\| \leq k^{1/2}$, (11.14) implies

$$e_k(\widetilde{A}) \leq c_3 k^{1/2}.$$

Therefore, we have by (11.13)

$$e_{2k}(u_{|H}\widetilde{A}) \leq e_k(u_{|H})e_k(\widetilde{A})$$
$$\leq c_2 c_3 \ell(u).$$

We turn now to the operator $v = Bu_{|H}\widetilde{A} : \ell_\infty^n \to \ell_\infty^k$.

Let $\lambda = c_2 c_3 \ell(u)$. Let $K = B_{\ell_\infty^n}$. The last estimate implies that $u_{|H}\widetilde{A}(K)$ is covered by 2^{2k} translates of λB_F. Therefore

(11.15) $\mathrm{vol}(Bu_{|H}\widetilde{A}(K))^{1/k} \leq 2^2 \lambda \, \mathrm{vol}(B(B_F))^{1/k},$

hence by (11.11)

$$\leq 2^2 C_1 k^{-1/2} \, \mathrm{vol}(K)^{1/k}.$$

Finally, since $v = Bu_{|H}\widetilde{A}$ satisfies

$$|\det(v)| = \frac{\mathrm{vol}(v(K))}{\mathrm{vol}(K)} \quad \text{and} \quad \det(v) = \det(< y_i^*, ux_j >),$$

(11.15) implies the above claim (11.12). This completes the proof that (c) $\Rightarrow$ (a). ∎

The *dual form* of Theorem 11.12 is the following result from [Pa1].

Theorem 11.13. *The following properties of a Banach space X are equivalent.*

(i) *X is a weak type 2 space.*

(vi) *There is a constant C such that for all n and all $w : \ell_1^n \to X$ we have*

(11.16) $$e_{[Cn]}(w) \leq C\|w\|n^{-1/2}$$

(vii) *There is a constant C such that for all n and all $w : \ell_1^n \to X$ we have*

(11.17) $$\left(\frac{\text{vol}(w(B_{\ell_1^n}))}{\text{vol}(B_{\ell_1^n})}\right)^{1/n} \leq C\|w\|n^{-1/2}.$$

Proof: (i) $\Rightarrow$ (vi). Assume (i). Let w be as in (vi). We can assume w of rank n. Let $E = w(\ell_1^n)$. Let us denote by $\overline{w} : \ell_1^n \to E$ the restriction (relative to the range) of w. Since by Proposition 11.4 we have $wC_2(E^*) \leq wT_2(E) \leq wT_2(X)$, the preceding result yields $e_{[Cn]}(\overline{w}^*) \leq C\|w\|n^{-1/2}$. By Corollary 8.11 this gives $e_{[C'n]}(\overline{w}) \leq C'\|w\|n^{-1/2}$ for some constant C'. Since $e_{[C'n]}(\overline{w})$ is equivalent to $e_{[C'n]}(w)$, (11.16) follows. This proves (i) $\Rightarrow$ (vi). (vi) $\Rightarrow$ (vii) is obvious

Finally, we prove (vii) $\Rightarrow$ (i). Assume (11.17). Then clearly X cannot contain ℓ_1^n's uniformly, hence by Theorem 11.3, X is K-convex. Moreover, using the (reverse) Santaló's inequality (Corollary 7.2 or Theorem 8.7) it is easy to see that (11.17) implies that X^* satisfies the property (c) in Theorem 11.12. Therefore, X^* is weak cotype 2. By Proposition 11.4, we conclude that X is a weak type 2 space. ∎

Finally, we come to the main result of [Pa1] which is the analogue for weak type 2 of the volume ratio characterization of weak cotype 2 obtained in Theorem 10.4.

Theorem 11.14. *The following properties of a Banach space X are equivalent.*

(i) *X is a weak type 2 space.*

(viii) *There is a constant C such that for all f.d. quotient spaces Q of X^* we have*
$$vr(Q) \leq C.$$

(ix) *There is a constant C such that for all n and all n-dimensional subspaces $E \subset X$ the minimal volume ellipsoid $D_E^{\min}$ containing B_E satisfies*

$$\left(\frac{\text{vol}(D_E^{\min})}{\text{vol}(B_E)} \right)^{1/n} \leq C.$$

Proof: The equivalence of (viii) and (ix) is obvious using Santaló's inequality and its converse (cf. Theorem 8.7). The implication (i) $\Rightarrow$ (viii) is easy using $wC_2(Q) \leq wT_2(Q^*) \leq wT_2(X)$ and Theorem 10.4 with the remark following it.

Finally, we show that (ix) implies (i). Assume (ix). We will show that (vi) holds. Consider $w : \ell_1^n \to E$ with $E \subset X$ $\dim E = n$. Let $v : E \to \ell_2^n$ be such that $\|v\| \leq 1$ and $D_E^{\min} = v^{-1}(B_{\ell_2^n})$. Consider the operator $T = vw : \ell_1^n \to \ell_2^n$. By Theorem 11.12 applied to $X = \ell_2$ (see also the following remark) we have

$$(11.18) \qquad e_{[C_1 n]}(T) \leq C_1 n^{-1/2} \|T\|$$

for some numerical constant C_1.

On the other hand, (ix) implies (by Lemma 7.5) that $N(D_E^{\min}, B_E) \leq (3C)^n$. Hence, if $c_2 = [\text{Log}_2 \, 3c] + 1$ we have

$$(11.19) \qquad e_{c_2 n}(v^{-1}) \leq 1.$$

Finally, writing $w = v^{-1}T$, we deduce from (10.18) and (11.19)

$$\begin{aligned} e_{[c_1 n] + c_2 n}(w) &\leq e_{c_2 n}(v^{-1}) e_{[c_1 n]}(T) \\ &\leq C_1 n^{-1/2} \|T\| \\ &\leq C_1 n^{-1/2} \|w\|. \end{aligned}$$

This shows that (vi) holds, and this concludes the proof of Theorem 11.14. ∎

Remark: The inequality (11.18) valid for any $T : \ell_1^n \to \ell_2^n$ is actually a variant of the classical Hadamard inequality for an $n \times n$ matrix (a_{ij})

$$|\det(a_{ij})|^{1/n} \leq \sup_j \left(\sum_i |a_{ij}|^2 \right)^{1/2}.$$

Indeed, this implies, if a_{ij} is the matrix associated to $T : \ell_1^n \to \ell_2^n$,

$$\left(\frac{\operatorname{vol}(TB_{\ell_1^n})}{\operatorname{vol}(B_{\ell_1^n})}\right)^{1/n} = |\det(a_{ij})|^{1/n} \leq \sup_j \|Te_j\|,$$

and equivalently

$$\left(\frac{\operatorname{vol}(TB_{\ell_1^n})}{\operatorname{vol}(B_{\ell_2^n})}\right)^{1/n} \leq \lambda(n)\|T\|,$$

with

$$\lambda(n) = (\operatorname{vol}(B_{\ell_1^n}))^{1/n}(\operatorname{vol}(B_{\ell_2^n}))^{-1/2} \sim Cn^{-1/2}.$$

The latter inequality can be used to prove (ix) $\Rightarrow$ (vii) in the preceding proof.

We will use the following result about K-convexity, which seems to be of independent interest.

Theorem 11.15. *A Banach space X is K-convex iff there is a constant C such that, for all n and for all n-dimensional subspaces $E \subset X$, there is a projection $P : X \to E$ such that $e_n(P) \leq C$.*

Proof: Assume X K-convex and consider $E \subset X$ with $\dim E = n$. By Corollary 3.12, there is an isomorphism $u : \ell_2^n \to E$ and an operator $v : X \to \ell_2^n$ such that $v_{|E} = u^{-1}$ and $\ell(u) = \ell^*(v) = n^{1/2}$. Since X is K-convex, we have $\ell(v) \leq K(X)n^{1/2}$ (by the remark after Lemma 3.10) hence by Theorem 5.8 we have

$$e_{\left[\frac{n}{2}\right]}(v) \leq K \text{ and } e_{\left[\frac{n}{2}\right]}(u) \leq K$$

for some constant K. Let $P = uv$. Clearly P is a projection from X onto E and

$$e_n(P) \leq e_{\left[\frac{n}{2}\right]}(u)e_{\left[\frac{n}{2}\right]}(v) \leq K^2.$$

This proves the only if part.

Conversely, assume that X satisfies the property considered in Theorem 11.15. By Theorem 11.3, it is enough to show that X does not contain ℓ_1^n's uniformly. Suppose on the contrary that X contains ℓ_1^n's uniformly. Then the property considered in Theorem 11.15 holds for $X = \ell_1^n$, uniformly with respect to n. We will show that this leads to a contradiction.

Let $D_n = \{-1, 1\}^n$ equipped with its uniform probability measure μ. Let $X = L_1(D_n, \mu)$ and let E be the span of the n-coordinate functions on D_n which we denote by $\varepsilon_1, \ldots, \varepsilon_n$. It is well known (by Khintchine's inequality, cf. [LT1], p. 66) that E is uniformly isomorphic to ℓ_2^n. Precisely, there is a positive numerical constant A_1 and an isomorphism

$$S : \ell_2^n \to E$$

such that

$$\|S\| \leq 1, \quad \|S^{-1}\| \leq (A_1)^{-1}.$$

Then we claim that for any projection $P : X \to E$ we have $e_n(P) \geq cn^{1/2}(\log n)^{-1}$ for all $n > 1$, for some numerical constant $c > 0$. This will be the announced contradiction. To prove this claim, let $Q = X/\operatorname{Ker} P$. Let $\sigma : X \to Q$ be the quotient map and let $T = \sigma_{|E} : E \to Q$. Note that there is an operator $\widetilde{P} : Q \to E$ (canonically associated to P) such that $\widetilde{P}\sigma = P$ and hence $e_n(P) = e_n(\widetilde{P})$. Since $P_{|E} = I_E$ we have $\widetilde{P}T = I_E$, which means that $\widetilde{P} = T^{-1}$. Therefore, to show the above claim, it suffices to bound $e_{[n/2]}(T)$ from above.

For that purpose we consider the operator $TS : \ell_2^n \to Q$. Let $i : E \to X$ be the inclusion mapping. We have $TS = \sigma iS$. We will show

(11.20) $$\ell^*((TS)^*) \leq (\pi/2)^{1/2}.$$

Indeed, this follows immediately from

(11.21) $$\ell^*((iS)^*) \leq (\pi/2)^{1/2}.$$

Let us check (11.21). Consider an arbitrary operator $u : \ell_2^n \to L_\infty(D_n)$. Let $\varphi_i = u(e_i)$. We have

$$\ell(u) = \left\| \sum g_i \varphi_i \right\|_{L_2(L_\infty(D_n))}.$$

We may assume (g_i) independent of (ε_i) and we set $\varepsilon_i' = \operatorname{sign}(g_i)$. Let then

$$\psi(\varepsilon, \varepsilon') = \prod_1^n (1 + \varepsilon_i \varepsilon_i').$$

We have (since $E|g_i| = (2/\pi)^{1/2}$)

$$(2/\pi)^{1/2} \sum <\varphi_i, \varepsilon_i> = \int \left(\sum g_i \varphi_i \right) \psi \, dP \, d\mu$$

$$\leq \left\| \sum g_i \varphi \right\|_{L_1(L_\infty(D_n))} \|\psi\|_{L_\infty(L_1(D_n))}$$

$$\leq \ell(u).$$

Hence

$$tr((iS)^* u) \leq (\pi/2)^{1/2} \ell(u).$$

This implies (11.21) and *a fortiori* (11.20). But since $(TS)^*$ is defined on the n-dimensional space Q^*, we have by Lemma 3.10 and Theorem 2.5

$$\ell(TS) \leq K(Q)\ell^*((TS)^*)$$
$$\leq K(1 + \text{Log } n)(\pi/2)^{1/2}.$$

Therefore, by Theorem 5.8, for some constant K' we have

$$e_n(TS) \leq K'n^{-1/2}(1 + \text{Log } n);$$

hence

(11.22)
$$e_n(T) \leq \|S^{-1}\|e_n(TS)$$
$$\leq (A_1)^{-1}K'n^{-1/2}(1 + \text{Log } n).$$

Since $\widetilde{P}T = I_E$, we have by (7.5)

$$\text{vol}(B_E) \leq 2^{2n-1}e_{2n}(I_E)^n \text{vol}(B_E);$$

hence

$$2^{-2} \leq e_{2n}(I_E) \leq e_n(\widetilde{P})e_n(T),$$

which implies by (11.22)

$$e_n(\widetilde{P}) \geq c \ n^{1/2}(\text{Log } n)^{-1}$$

for some numerical constant $c > 0$. This proves the above claim and completes the proof. ■

Remark: *A fortiori* if X is K-convex there is a constant C such that for all n and all n-dimensional subspaces $E \subset X$, there is a projection $P : X \to E$ such that

$$\left(\frac{\text{vol}(P(B_X))}{\text{vol}(B_E)}\right)^{1/n} \leq 2C.$$

Conversely, the preceding proof shows that this property also characterizes K-convex spaces.

We will use the following reformulation of Theorem 11.15.

Lemma 11.16. *Let X be a Banach space. For a subspace $E \subset X$, let $i_E : E \to X$ be the inclusion map. Similarly, for a quotient space Q of X, let $\sigma_Q : X \to Q$ denote the quotient map. Let us denote by T_{EQ} the composition $T_{EQ} = \sigma_Q i_E$. Note that $\|T_{EQ}\| \leq 1$. Then, if X is K-convex, there is a constant C such that for all n, for each n-dimensional subspace E, there is an n-dimensional quotient space Q such that T_{EQ} is invertible and*

$$(11.23) \qquad e_n((T_{EQ})^{-1}) \leq C.$$

A fortiori we have

$$(11.24) \qquad \mathrm{vol}(B_Q)^{1/n} \leq 2C \, \mathrm{vol}(\sigma_Q(B_E))^{1/n}.$$

Proof: By Theorem 11.15 there is a projection $P : X \to E$ such that $e_n(P) \leq C$. Let $Q = X/\operatorname{Ker} P$ and let $\widetilde{P} : Q \to E$ be such that $\widetilde{P}\sigma_Q = P$. Clearly $\widetilde{P} = (T_{EQ})^{-1}$ and $e_n(P) = e_n(\widetilde{P})$ so that (11.23) follows. *A fortiori* we obtain (11.24) since $\sigma_Q(T_{EQ})^{-1}(B_Q) \supset B_Q$. ∎

With the help of this lemma, we may slightly strengthen the property (ix) in Theorem 11.14 as follows.

Theorem 11.17. *Let X be a weak type 2 space. Then there is a constant C' such that for all n and all n-dimensional subspaces $E \subset X$ there is an operator $v : X \to \ell_2^n$ with $\|v\| \leq 1$ such that*

$$(11.25) \qquad \left(\frac{\mathrm{vol}((v_{|E})^{-1}(B_{\ell_2^n}))}{\mathrm{vol}(B_E)} \right)^{1/n} \leq C'.$$

Proof: Let $E \subset X$ be of dimension n. By the preceding lemma, there is a quotient Q of X satisfying (11.24).

Clearly $wT_2(Q) \leq wT_2(X)$, so that by (ix) there is a constant C_1 (depending only on $wT_2(X)$) such that

$$\left(\frac{\mathrm{vol}(D_Q^{\min})}{\mathrm{vol}(B_Q)} \right)^{1/n} \leq C_1.$$

Thus, by (11.24) we have

(11.26) $$\left(\frac{\mathrm{vol}(D_Q^{\min})}{\mathrm{vol}(T_{EQ}(B_E))} \right)^{1/n} \leq 2CC_1.$$

Now, assume $D_Q^{\min} = u^{-1}(B_{\ell_2^n})$ for some $u : Q \to \ell_2^n$ with $\|u\| \leq 1$. Then we can define $v = u\sigma_Q : X \to \ell_2^n$. We have $\|v\| \leq \|u\| \leq 1$ and $v_{|E}(B_E) = u\,T_{EQ}(B_E)$. This shows that $v_{|E}$ is an isomorphism and $B_E = (v_{|E})^{-1}u\,T_{EQ}(B_E)$. Thus (11.26) implies

$$\left(\frac{\mathrm{vol}((v_{|E})^{-1}uD_Q^{\min})}{\mathrm{vol}((v_{|E})^{-1}uT_{EQ}(B_E))} \right)^{1/n} \leq 2CC_1,$$

from which (11.25) follows with $C' = 2CC_1$. ∎

Remark: By an obvious modification, we can obtain $e_n((v_{|E})^{-1}) \leq C'$ for some constant C'.

Remark: This refines Theorem 11.14 since we find an ellipsoid $D = (v_{|E})^{-1}(B_{\ell_2^n})$ which *not only* satisfies $B_E \subset D$ but *also* there is a linear projection $P : X \to E$ ($P = (v_{|E})^{-1}\circ v$) such that $P(B_X) \subset D$. In other words, we have proved the following.

Corollary 11.18. *Let X be a weak type 2 space. Then there is a constant C such that for all n and all n-dimensional subspaces $E \subset X$ there is a Euclidean norm $|\ |$ on E and a linear projection $P : X \to E$ such that $\forall\, x \in E \ |Px| \leq \|x\|$ and if $D = \{x \in E|\ |x| \leq 1\}$ we have*

$$\left(\frac{\mathrm{vol}(D)}{\mathrm{vol}(B_E)} \right)^{1/n} \leq C.$$

Obviously each of the properties considered in Theorem 11.17 and its corollary characterizes weak type 2 spaces.

Notes and Remarks

The first part of this chapter is based on [MP1]. Except Theorem 11.3 (from [P3]) and the elementary Lemma 11.7, all the statements from Definition 11.1 to Theorem 11.10 are essentially in [MP1]. Occasionally, we have included a slight improvement over [MP1], for instance the equivalence between (i) and (iv) which has been observed independently by N. Tomczak-Jaegermann. For a refinement of Lemma 11.7, cf. [P6]. Lemma 11.11 comes from [Ca2]. (It is implicit in [Pi2].)

The *volume ratio characterization* of weak type 2 spaces in Theorems 11.13 and 11.14 is due to A. Pajor [Pa1] as well as the main implication (c) $\Rightarrow$ (a) in Theorem 11.12 ((a) $\Rightarrow$ (b) is in [MP1] and (b) $\Rightarrow$ (c) is essentially obvious).

As far we know, Theorem 11.15, Lemma 11.16 and Theorem 11.17 and its Corollary are new, but they might very well have been observed by others. In particular, the *only if* part of Theorem 11.15 is merely a combination of known results.

Chapter 12

Weak Hilbert Spaces

By a well-known result of Kwapień [K1], a Banach space is both of type 2 and cotype 2 iff it is isomorphic to a Hilbert space. We are thus led naturally to the following:

Definition 12.1. *A Banach space X is called a weak Hilbert space if it is both weak type 2 and weak cotype 2.*

By the previous results, it is clear that if X is weak Hilbert then all subspaces and all quotient spaces of X are also weak Hilbert. Also, every space Y which is finitely representable in X is a weak Hilbert space. Moreover, X is weak Hilbert iff its dual X^* is weak Hilbert (this follows from Proposition 11.4). Moreover, X is a weak Hilbert space iff it is K-convex and X, X^* are both weak cotype 2. This notion is clearly an *isomorphic* property; we will see later that there is apparently no isometric analogue.

We recall that we say that a subspace $F \subset X$ is C-complemented in X if there is a projection $P : X \to F$ with $\|P\| \leq C$. Let X, Y be Banach spaces and let $T : X \to Y$ be an operator which factors through a Hilbert space H. We recall that the *norm of factorization through Hilbert space* of T is denoted by $\gamma_2(T)$ and is defined as

$$\gamma_2(T) = \inf\{\|A\| \cdot \|B\|\},$$

where the infimum runs over all Hilbert spaces H and all operators $B : X \to H$ and $A : H \to Y$ such that $T = AB$.

Let $F \subset X$ be a subspace of an arbitrary space X. Assume that there is a projection $P : X \to F$ such that $\gamma_2(P) \leq \lambda$. Then clearly this implies

$$(12.1) \qquad d_F \leq \lambda \text{ and } \|P\| \leq \lambda.$$

We also recall that for all $u : \ell_2^n \to X$ we have by (3.14) $\ell(u) \leq \pi_2(u)$. This implies by duality, for any $v : X \to \ell_2^n$,

$$(12.2) \qquad \pi_2(v) = \pi_2^*(v) \leq \ell^*(v).$$

189

In this chapter we give twelve equivalent reformulations of the notion of a weak Hilbert space. They will be numbered from (i) to (xii). We start with a first series of six properties.

Theorem 12.2. *The following properties of a Banach space X are equivalent*

(i) X *is a weak Hilbert space.*

(ii) *For every $0 < \delta < 1$, there is a constant $\widetilde{C}_\delta$ with the following property: every f.d. subspace $E \subset X$ contains a subspace $F \subset E$ with $\dim F \geq \delta \dim E$ such that $d_F \leq \widetilde{C}_\delta$ and there is a projection $P : X \to F$ with $\|P\| \leq \widetilde{C}_\delta$.*

(iii) *There is $0 < \delta_0 < 1$ and a constant $\widetilde{C}_{\delta_0}$ such that the same as (ii) holds.*

(iv) *There are constants C and $0 < \alpha < 1$ such that for all n and all $u : \ell_2^n \to X$ we have*

$$a_{[\alpha n]}(u) \leq C([\alpha n])^{-1/2} \pi_2(u^*).$$

(v) *There is a constant C such that, for any Hilbert space H, every operator $u : H \to X$ with a 2-summing adjoint satisfies*

$$\sup k^{1/2} a_k(u) \leq C \; \pi_2(u^*).$$

(vi) *There is a constant C such that for all n and all n-dimensional subspaces $E \subset X^*$ we can find for each $k = 1, 2, \ldots, n$ a subspace $F \subset E^*$ with $\dim F > n - k$ and a projection $P : X^* \to F$ satisfying*

$$\gamma_2(P) \leq C(n/k)^{1/2}.$$

Proof: (i) $\Rightarrow$ (ii) This follows immediately from Theorem 10.2 and Theorem 11.6.

(ii) $\Rightarrow$ (iii) is trivial.

Let us show that (iii) $\Rightarrow$ (iv). Assume (iii). Consider $u : \ell_2^n \to X$. We may assume without loss of generality that $E = u(\ell_2^n)$ is of dimension n. Let $\alpha = 1 - \delta_0/2$. Applying (iii) to E we find a projection $P : X \to F$ with $F \subset E$ satisfying $\dim F = [\delta_0 n] + 1$ and $\gamma_2(P) \leq (\widetilde{C}_{\delta_0})^2$. Let $K = (\widetilde{C}_{\delta_0})^2$.

We claim that

$$\left(\sum a_k(Pu)^2\right)^{1/2} \leq \gamma_2(P)\pi_2(u^*).$$

Indeed, let $P = AB$ with $B : X \to H, A : H \to F$.

$$H$$
$$B \nearrow \qquad \searrow A$$
$$\ell_2^n \xrightarrow{u} X \xrightarrow{P} F$$

Then Bu is an operator between Hilbert spaces; hence by (1.13) and Proposition 1.6 we have

$$\left(\sum a_k(Bu)^2\right)^{1/2} = \pi_2(Bu) = \pi_2((Bu)^*) \le \|B\|\pi_2(u^*).$$

Therefore

$$\left(\sum a_k(ABu)^2\right)^{1/2} \le \|A\| \, \|B\|\pi_2(u^*).$$

Taking the infimum over all factorizations $P = AB$, we obtain the above claim. This claim implies

(12.3) $$a_k(Pu) \le K \, \pi_2(u^*)k^{-1/2}.$$

Since the rank of $u - Pu$ is $n - [\delta_0 n] - 1$, choosing $k = [\delta_0 n/2]$ in (12.3) we obtain

$$a_{[\alpha n]}(u) \le K \, \pi_2(u^*)(2/\delta_0 n)^{1/2},$$

with $\alpha = 1 - \delta_0/2$. This yields (iv).

The implication (iv) $\Rightarrow$ (v) is an immediate consequence of Lemma 10.3. Let us prove (v) $\Rightarrow$ (vi). Assume (v).

Let $E \subset X^*$ be a subspace with $\dim E = n$. By Theorem 3.8 we know that there is an isomorphism $T : \ell_2^n \to E$ such that $\|T\| = 1$ and $\pi_2(T^{-1}) = n^{1/2}$. Actually (by Hahn–Banach), for each $\varepsilon > 0$, T^{-1} admits an extension $v : X^* \to \ell_2^n$ which is weak $*$-continuous and satisfies $v_{|E} = T^{-1}$, $\pi_2(v) \le n^{1/2}(1 + \varepsilon)$. Then we can write $v = u^*$ for some $u : \ell_2^n \to X$ and (v) implies $\sup k^{1/2}a_k(v) \le C \, n^{1/2}(1 + \varepsilon)$. Therefore, by definition of $a_k(v)$, there is a projection $Q : \ell_2^n \to \ell_2^n$ with rank $> n - k$, such that

$$\|Qv\| \le C(n/k)^{1/2}(1 + \varepsilon).$$

Let F be the range of TQ. Then $\dim F > n - k$ and $P = TQv$ is clearly a projection onto F. Moreover,

$$\gamma_2(P) \le \|T\| \, \|Qv\|$$
$$\le C(n/k)^{1/2}(1 + \varepsilon).$$

This proves that (vi) holds.

Let us record here that by Theorem 10.2 (vi) clearly implies that X^* is weak cotype 2. Let us show first that (v) $\Rightarrow$ (i). Assume (v). Then, for any $u : \ell_2^n \to X$, we have by (12.2)

$$\sup k^{1/2} a_k(u) \le C \, \pi_2(u^*) \le C \, \ell^*(u^*),$$

and this immediately yields that X^* is a weak type 2 space. Since (v) $\Rightarrow$ (vi) we know by the observation recorded above that X^* is weak cotype 2, so we conclude that X^* is a weak Hilbert space, hence X is a weak Hilbert space. This shows that properties (i) to (v) are equivalent. Finally, let us show that (vi) $\Rightarrow$ (i). Assume (vi). Then X^* satisfies (ii). Hence X^* is weak Hilbert and therefore X is weak Hilbert also. This shows that (vi) $\Rightarrow$ (i) and concludes the proof of Theorem 12.1.

Remark: *A posteriori* we may replace X^* by X in all the properties listed in Theorem 12.2, in particular in (vi).

We collect now a number of facts which follow from the previous two chapters.

Theorem 12.3.   *Let X be a weak Hilbert space.*

(1) X *is of type p and of cotype q for all $p < 2$ and all $q > 2$.*

(2) *Let $x_1, \ldots, x_n$ be a normalized λ-unconditional sequence in X. Then we have, for all (α_i) in $\mathbf{R}^n$,*

$$\frac{1}{C\lambda^2}\|\alpha\|_{2\infty} \le \left\|\sum \alpha_i x_i\right\| \le C \, \lambda^2 \|\alpha\|_{21}$$

for some constant C depending only on $wC_2(X)$ and $wT_2(X)$. Moreover, this extends to an infinite normalized λ-unconditional sequence $\{x_n | n \ge 1\}$.

(3) X *is isomorphic to a Hilbert space iff one of the spaces $\ell_2(X)$, $L_2(X), \mathrm{Rad}(X)$ or $G(X)$ is a weak Hilbert space.*

Proof:

(1) Follows from Proposition 10.7 and Proposition 11.2.

(2) Follows from Corollary 10.9 and Proposition 11.9.

(3) Follows from Proposition 10.11 and Remark 11.8.

As will be seen in the next chapter, the only known examples of weak Hilbert spaces have an unconditional basis. It is therefore of interest to investigate further what happens in that case. The next result is a reformulation of an unpublished lemma of Gordon–Lewis.

Proposition 12.4. *Let $x_1, \ldots, x_n$ be a normalized λ-unconditional sequence in X. Then, if X is weak-Hilbert for all $0 < \delta < 1$, we can find a subset $A \subset \{1, \ldots, n\}$ with $|A| \geq \delta n$ such that $(x_i)_{i \in A}$ is $f(\lambda, \delta)$-equivalent to $\ell_2^{|A|}$ where $f(\lambda, \delta)$ depends only on X, λ, δ.*

Proof: Let E be the span of $x_1, \ldots, x_n$. Since X is weak-Hilbert, there is an operator $T : E \to E$ with $\mathrm{rk}(T) \geq n/2$ and $\gamma_2(T) \leq C_1$ for some constant C_1 depending only on X. Let $G = \{-1, 1\}^n$ equipped with its uniform probability measure and let $T_g : E \to E$ be the operator defined by $T g\, x_i = g_i x_i$. Then let

$$\widetilde{T} = \int T_g^{-1} T\, T_g \; d\mu(g).$$

This is clearly a *diagonal* operator on E with respect to $x_1, \ldots, x_n$ and we have

$$\gamma_2(\widetilde{T}) \leq C_1.$$

Moreover, $\mathrm{tr}\, \widetilde{T} = \mathrm{tr}\, T \geq n/2$. Therefore, the diagonal coefficients $\lambda_1, \ldots, \lambda_n$ of $\widetilde{T}$ satisfy $\sum \lambda_i \geq n/2$ and (since $\|\widetilde{T}\| \leq \gamma_2(\widetilde{T}) \leq C_1$) $|\lambda_i| \leq C_1$. This implies that if $A = \{i \mid |\lambda_i| \geq 1/4C_1\}$ then $|A| \geq n/4C_1$ and we have $\forall \; (\alpha_i)_{i \in A} \in \mathbb{R}^A$

$$\frac{1}{\lambda} \frac{1}{4C_1} \left\| \sum \alpha_i x_i \right\| \leq \left\| \sum \alpha_i \lambda_i x_i \right\| \leq C_1 \left\| \sum \alpha_i x_i \right\|.$$

But since $\gamma_2(\widetilde{T}) \leq C_1$ and $\|\widetilde{T}\alpha\| = \|\sum \alpha_i \lambda_i x_i\|$ we can find elements h_i in a Hilbert space H such that

$$(12.4) \qquad \left\| \sum \alpha_i \lambda_i x_i \right\| \leq \left\| \sum \alpha_i h_i \right\|_H \leq C_1 \left\| \sum \alpha_i x_i \right\|,$$

and averaging over $\pm \alpha_i$ we obtain

$$(12.5) \qquad \frac{1}{\lambda} \left\| \sum \alpha_i \lambda_i x_i \right\| \leq \left(\sum |\alpha_i|^2 \|h_i\|^2 \right)^{1/2} \leq C_1 \lambda \left\| \sum \alpha_i x_i \right\|.$$

Note that by (12.4) we have

$$\frac{1}{4C_1} \leq \|h_i\| \leq C_1;$$

hence (12.5) implies

$$\frac{1}{4C_1^2 \lambda} \left(\sum_{i \in A} |\alpha_i|^2 \right)^{1/2} \leq \left\| \sum_{i \in A} \alpha_i x_i \right\| \leq 4C_1 \lambda^2 \left(\sum_{i \in A} |\alpha_i|^2 \right)^{1/2}.$$

Thus we obtain the conclusion with $|A| \geq n/(4C_1)$. Since $(x_1, \ldots, x_n)$ is λ-unconditional, we can clearly reiterate this argument for the complement of A (and so on) if we wish to increase $|A|$ to $|A| \geq \delta n$. ∎

Remark: Let X be a weak Hilbert space possessing an unconditional basis. Assume that X satisfies the conclusion of Proposition 12.4 (perhaps only for blocks of the basis of X), it seems likely that X should then be a weak Hilbert space but we do not know this. (Actually, such a converse might even be true in general. Compare with the discussion of property $\overset{\circ}{H}$ in the later Chapter 14.)

Recall that we denote by $D_E^{\max}$ (resp. $D_E^{\min}$) the maximal (resp. minimal) volume ellipsoid contained in (resp. containing) B_E.

We have the following volume ratio characterization of weak Hilbert spaces.

Theorem 12.5.   *The following properties of a Banach space X are equivalent.*

(i) *X is a weak Hilbert space.*

(vii) *There is a constant C such that for all n and all $E \subset X$ with $\dim E = n$ we have*

$$(12.6) \qquad \left(\frac{\mathrm{vol}(D_E^{\min})}{\mathrm{vol}(D_E^{\max})} \right)^{1/n} \leq C$$

(recall $D_E^{\max} \subset B_E \subset D_E^{\min}$).

(viii) *There is a constant C such that for all even integers n, every n-dimensional subspace $E \subset X$ admits a decomposition $E = E_1 + E_2$ with $\dim E_1 = \dim E_2 = n/2$ such that $d_{E_1} \leq C, d_{E_2} \leq C$ and there are projections $Q_1 : X \to E_1$ and $Q_2 : X \to E_2$ with norms*

$$\|Q_1\| \leq C, \quad \|Q_2\| \leq C.$$

Proof: The equivalence (i) $\Leftrightarrow$ (vii) is an immediate consequence of Theorem 10.4 and Theorem 11.14.

Let us show (i) $\Rightarrow$ (viii). Actually, we give two proofs of this. (The second proof is rather straightforward.) Assume (i). Then by Theorem 10.4 and Corollary 11.17, there are constants C_1, C_2 such that for any n-dimensional $E \subset X$, we can find two Euclidean norms $|\ |_1$ and $|\ |_2$ on E with associated unit balls (= ellipsoids) D_1, D_2 such that $D_1 \subset B_E \subset D_2$

$$\left(\frac{\mathrm{vol}(B_E)}{\mathrm{vol}(D_1)}\right)^{1/n} \leq C_1 \text{ and } \left(\frac{\mathrm{vol}(D_2)}{\mathrm{vol}(B_E)}\right)^{1/n} \leq C_2,$$

and, moreover, there is a projection $P : X \to E$ such that

(12.7) $$|Px|_2 \leq \|x\| \quad \forall\, x \in X.$$

A fortiori we have

$$\left(\frac{\mathrm{vol}(D_2)}{\mathrm{vol}(D_1)}\right)^{1/n} \leq C_1 C_2.$$

By Corollary 6.3 (applied to the inclusion $D_1 \subset D_2$!) there is a decomposition $E = E_1 + E_2$ with $\dim E_1 = \dim E_2 = n/2$ such that $\forall\, x \in E_1 \cup E_2$

(12.8) $$K^{-1}|x|_1 \leq |x|_2 \leq |x|_1,$$

where K is a fixed constant ($K = 4\pi C_1 C_2$). *A fortiori* we have

(12.9) $$K^{-1}|x|_1 \leq \|x\| \leq |x|_1 \quad \forall\, x \in E_1 \cup E_2.$$

Moreover, let P_1, P_2 be the orthogonal projections—relative to the scalar product induced by $|\ |_2$—onto E_1 and E_2 respectively. We have by (12.8) and (12.9), $\forall\, x \in X$

$$\|P_1 Px\| \leq |P_1 Px|_1 \leq K|P_1 Px|_2 \leq K|Px|_2;$$

hence by (12.7)

$$\leq K\|x\|.$$

This shows that $Q_1 = P_1 P$ is a projection onto E_1 with $\|Q_1\| \leq K$. Similarly, for $Q_2 = P_2 P$.

This shows that (i) $\Rightarrow$ (viii). There is a much more direct proof of this as follows. We use the implication (i) $\Rightarrow$ (ii) from Theorem 12.2 and take $\delta = 1/2$ in (ii). This gives us E_1. Now we can take for E_2 any space which is spanned by a small enough perturbation of a basis of E_1 and which satisfies $E_1 \cap E_2 = \{0\}$. By well-known properties of such perturbations E_2 will have the same properties as E_1 so that (viii) holds. Note, however, that the first proof gives a bit more information than stated (namely E_1, E_2 are orthogonal with respect to the Euclidean structure associated to D_1). Conversely, it is clear that (viii) implies the second property in Theorem 12.2; this completes the proof. ∎

Remark: It is rather surprising at first glance that (viii) can hold without E being uniformly isomorphic to $E_1 \oplus E_2$. A close inspection shows that E_2 has no reason to lie in the kernel of Q_1 in general, so we cannot deduce from (viii) a *good* decomposition of E.

Note that if we strengthen (12.6) to

$$D_E^{\min} \subset CD_E^{\max},$$

then of course X must be isomorphic to a Hilbert space.

Remark: By Proposition 11.4 and Theorem 10.4, a Banach space X is weak Hilbert iff it is K-convex and there is a constant C such that

$$\forall\, E \subset X \quad \forall\, F \subset X^* \quad vr(E) \leq C\ vr(F) \leq C.$$

The definition of weak Hilbert spaces given above is not entirely satisfactory. In particular, it does not yield a good notion of **weak Hilbert constant**. This will be provided by the properties considered in the next statement.

We need to recall some background first. Let X, Y be Banach spaces. Recall that an operator $T : X \to Y$ is called nuclear if there are two sequences $(x_n^*$ and $y_n)$ in X^* and Y respectively such that $\sum \|y_n\|\ \|x_n^*\| < \infty$ and

$$(12.10) \qquad Tx = \sum_{n=1}^{\infty} x_n^*(x)y_n \quad \forall\, x \in X.$$

The nuclear norm is defined as

$$N(T) = \inf \left\{ \sum \|x_n^*\|\ \|y_n\| \right\},$$

where the infimum runs over all representations (12.10). We denote by $N(X, Y)$ the space of all such operators equipped with this norm. It is well known that if E is a *finite-dimensional* Banach space, we have isometrically $B(E, X)^* = N(X, E)$ and $N(E, X)^* = B(X, E)$. The duality here is of course relative to the trace, i.e. we set

$$(12.11) \qquad < v, u >= tr\ uv \text{ for } u : E \to X,\ v : X \to E.$$

We will also need to work with the norm γ_2^* which is dual to the γ_2-norm. We will denote by $\Gamma_2^*(X,Y)$ the space of all operators $T : X \to Y$ which, for some Hilbert space H, can be written as

$$(12.12) \qquad T = BA \quad \text{with} \quad A \in \Pi_2(X, H), \quad B^* \in \Pi_2(Y^*, H).$$

We let $\gamma_2^*(T) = \inf\{\pi_2(A)\pi_2(B^*)\}$ where the infimum runs over all such representations. It is known (if e.g. [P2] Chapter 2) that if $S : Y \to X$ is a finite rank operator then

$$(12.13) \qquad \gamma_2(S) = \sup\{tr(ST)\,|\,T \in \Gamma_2^*(X,Y), \gamma_2^*(T) \le q1\}.$$

Moreover, for a finite-dimensional Banach space E we have the isometric identities

$$\Gamma_2(E, X)^* = \Gamma_2^*(X, E) \quad \text{and} \quad \Gamma_2^*(E, X)^* = \Gamma_2(X, E).$$

The duality here is again as in (12.11). If X is isomorphic to a Hilbert space, then every T in $\Gamma_2^*(X, X)$ must be nuclear and have summable eigenvalues (since this is true in ℓ_2). Clearly, (12.13) implies that the converse is also true. Precisely, we have: A complex Banach space X is isomorphic to a Hilbert space iff there is a constant C such that the eigenvalues $(\lambda_n(T))$ of any finite rank operator $T : X \to X$ satisfy

$$(12.14) \qquad \sum |\lambda_n(T)| \le C\gamma_2^*(T).$$

This result can also be reformulated in a less technical manner as follows (cf. [K2]).

Let $\lambda \ge 1$ be a constant. A Banach space X is λ-isomorphic to a Hilbert space iff for all bounded operators $A = (a_{ij})$ on ℓ_2 we have for all finitely supported sequences $x = (x_n)$ in X

$$(12.15) \qquad \left(\sum_i \left\|\sum_j a_{ij}x_j\right\|^2\right)^{1/2} \le \lambda\|A\| \left(\sum \|x_j\|^2\right)^{1/2}.$$

Equivalently, (12.15) means that for all finitely supported sequences $x^* = (x_j^*)$ in X^* we have

$$(12.16) \qquad \left|\sum_{ij} a_{ij} <x_i^*, x_j>\right| \le \lambda\|A\| \left(\sum \|x_j\|^2 \sum \|x_i^*\|^2\right)^{1/2}.$$

We denote by C_1 the space of all nuclear (i.e. *trace class*) operators on ℓ_2 equipped with the usual norm. Moreover, we denote by $M(x, x^*)$ the doubly infinite matrix $(< x_i^*, x_j >)$ considered as an operator on ℓ_2. Then, taking the supremum over all $A : \ell_2 \to \ell_2$ with $\|A\| \leq 1$ in (12.16), we obtain

$$(12.17) \qquad \|M(x, x^*)\|_{C_1} \leq \lambda \|x\|_{\ell_2(X)} \|x^*\|_{\ell_2(X^*)}.$$

We will denote by $C_{1\infty}$ the space of all compact operators $T : \ell_2 \to \ell_2$ such that

$$\sup_n n\lambda_n(|T|) < \infty$$

and we set

$$\|T\|_{1\infty} = \sup_{n \geq 1} n\lambda_n(|T|).$$

Clearly, $\|T\|_{1\infty} \leq \|T\|_1$.

In the next result we show that if we replace ℓ_1 by weak ℓ_1 and C_1 by $C_{1\infty}$ in (12.14) and (12.17) then we obtain a characterization of weak Hilbert spaces.

Theorem 12.6. *The following properties of a Banach space X are equivalent.*

(i) *X is a weak Hilbert space.*

(ix) *There is a constant C such that every operator T in $\Gamma_2^*(X, X)$ is compact and satisfies*

$$\sup n|\lambda_n(T)| \leq C\gamma_2^*(T).$$

(We recall that the sequence $\{|\lambda_n(T)|, n \geq 1\}$ is always assumed to be rearranged in non-increasing order.)

(x) *There is a constant C such that, for all x in $\ell_2(X)$ and all x^* in $\ell_2(X^*), M(x, x^*)$ belongs to $C_{1\infty}$ and satisfies*

$$\|M(x, x^*)\|_{1\infty} \leq C\|x\|_{\ell_2(X)} \|x^*\|_{\ell_2(X^*)}.$$

(xi) *There is a constant C such that, for all n, for all $x_1, \ldots, x_n$ in B_X and all $x_1^*, \ldots, x_n^*$ in B_{X^*} we have*

$$|\det(< x_i^*, x_j >)| \leq C^n.$$

(xii) *There is a constant C such that for all nuclear operators $T : X \to X$, the eigenvalues of T satisfy*

$$\sup_n n|\lambda_n(T)| \leq CN(T).$$

Note: We will denote by $w\gamma_2(X)$ the smallest constant C such that (x) holds. We feel that (x) should be taken as the *right* definition of a weak Hilbert space.

Remark: Whenever we are dealing with eigenvalues as in (ix) or (xii) above, we *implicitly* assume that the space X is a *complex* Banach space.

Note, however, that this does not really restrict the generality since most of the properties that we consider can be reduced to the complex case if we wish. Indeed, let X be a real Banach space. Consider its complexification $\widetilde{X}$ defined as the space of all $\mathbb{R}$-linear operators from $\mathbb{C}$ into X. As real Banach spaces $\widetilde{X} \approx X \oplus X$ and $X \oplus X$ is weak Hilbert iff X is weak Hilbert. Moreover, the definition of weak Hilbert is independent of the choice of scalars $\mathbb{R}$ or $\mathbb{C}$ so that $\widetilde{X}$ is a weak Hilbert space iff X is a weak Hilbert space. Similarly, each of the properties (x), (xi) above holds for X iff it holds for $\widetilde{X}$.

This remark allows to extend the equivalence of (i), (x), and (xi) to the real case.

Proof of Theorem 12.6: We will leave the proof of the (xi) $\Rightarrow$ (xii) for the next chapter. Let us show that (i) $\Rightarrow$ (ix). Assume (i). Consider T in $\Gamma_2^*(X, X)$ and a factorization $T = BA$ as in (12.12). By Theorem 12.2 (recalling that X^* is also weak Hilbert) there is a constant C such that

$$a_k(A^*) = a_k(A) \leq Ck^{-1/2}\pi_2(A)$$

and

$$a_k(B) \leq Ck^{-1/2}\pi_2(B^*)$$

for all k. This implies that $T = BA$ is compact and we have

$$a_{2k}(T) \leq a_k(A)a_k(B) \leq C^2 k^{-1}\pi_2(A)\pi_2(B^*)$$

so that we find

$$\sup_n na_n(T) \leq 4C^2\gamma_2^*(T).$$

By a known result (cf. e.g. [Pi2]) we have

$$\sup n|\lambda_n(T)| \leq K \sup na_n(T)$$

for some constant K. Therefore we obtain (ix). Let us prove (ix) $\Rightarrow$ (x). Assume (ix).

Consider $x \in \ell_2(X)$ and $x^* \in \ell_2(X^*)$. We define the operators $A : X \to \ell_2$ and $B^* : X^* \to \ell_2$ as follows:

$$\forall\, x \in X \quad Ax = (x_j^*(x))_{j \geq 1}$$
$$\forall\, x^* \in X^* \quad B^* x^* = (x^*(x_i))_{i \geq 1}.$$

Clearly B^* is the adjoint of an operator $B : \ell_2 \to X$. It is easy to check that

$$\pi_2(A) \leq \left(\sum \|x_j^*\|^2 \right)^{1/2}$$

and

$$\pi(B^*) \leq \left(\sum \|x_i\|^2 \right)^{1/2}.$$

Therefore, for any operator $U : \ell_2 \to \ell_2$, with norm 1, the composition $T = BUA$ belongs to $\Gamma_2^*(X, X)$ and

$$(12.18) \qquad \gamma_2^*(T) \leq \|x\|_{\ell_2(X)} \|x^*\|_{\ell_2(X^*)}.$$

By (ix) we have $\sup n|\lambda_n(BUA)| \leq C\gamma_2^*(T)$, but we observe that the non-zero eigenvalues of BUA are the same (with multiplicity) as those of UAB (cf. e.g. [Pi1]). Therefore (12.18) implies

$$(12.19) \qquad \sup n|\lambda_n(UAB)| \leq C\|x\|_{\ell_2(X)} \|x^*\|_{\ell_2(X^*)}.$$

It remains to choose U so that $UAB = |AB|$ (polar decomposition) and to observe that AB coincides with $M(x, x^*)$. We then deduce (x) from (12.19).

(x) $\Rightarrow$ (xi). Assume (x). Note that for any $n \times n$ matrix M we have by the polar decomposition

$$|\det(M)| = |\det(|M|)| = \prod_1^n \lambda_k(|M|);$$

hence $|\det(M)|^{1/n} \leq (\sup k\lambda_k(|M|)) \cdot (n!)^{-1/n}$. By Stirling's formula this implies

$$|\det(M)|^{1/n} \leq Kn^{-1} \sup_{1 \leq k \leq n} k\lambda_k(|M|)$$

for some numerical constant K.

Assuming (x), this implies for all x_i in X and all x_i^* in X^*

$$(12.20) \qquad |\det <x_i^*, x_j>|^{1/n} \leq Kn^{-1} \left(\sum_1^n \|x_i\|^2 \sum_1^n \|x_j^*\|^2 \right)^{1/2}$$

and (xi) follows immediately from (12.20). We will prove the implication (xi) $\Rightarrow$ (xii) in Chapter 15. We now prove (xii) $\Rightarrow$ (i).

Assume (xii). We will show that property (iii) (in Theorem 12.2) holds.

We proceed in several steps.

Step 1: Let F be an arbitrary f.d. quotient space of a subspace of X. We claim that F and F^* both satisfy (xii). To justify this claim, note that for $T : F \to F$ we have $N(T) = N(T^*)$; also T and T^* have the same eigenvalues (with multiplicity); hence it is enough to check that F satisfies (xii). For that purpose, we consider $X_2 \subset X_1 \subset X$ such that $F = X_1/X_2$. We denote by $\sigma : X_1 \to X_1/X_2$ the quotient mapping. Let $\varepsilon > 0$. For any $T : F \to F$, there is clearly a *lifting* $T_1 : F \to X_1$ such that $\sigma T_1 = T$ and $N(T_1) \leq (1 + \varepsilon)N(T)$. Let then $\widetilde{T}_1 = T_1 \sigma : X_1 \to X_1$. We have $N(\widetilde{T}_1) \leq N(T_1)$. By the Hahn–Banach Theorem, there is clearly an extension $\widetilde{T} : X \to X$ such that $N(\widetilde{T}) \leq N(\widetilde{T}_1)$ and $\widetilde{T}_{|X_1} = \widetilde{T}_1$. We have thus associated to T an operator $\widetilde{T}$ on X with $N(\widetilde{T}) \leq (1 + \varepsilon)N(T)$ such that the eigenvalues of T (with multiplicity) are included into those of $\widetilde{T}$. (Indeed, T and $\widetilde{T}_1$ are *related* since $\widetilde{T}_1 = T_1\sigma$ and $T = \sigma T_1$ and $\widetilde{T}$ extends $\widetilde{T}_1$.) Hence, assuming (xii), we find

$$\sup n|\lambda_n(T)| \leq \sup n|\lambda_n(\widetilde{T})| \leq (1 + \varepsilon)C\ N(T).$$

This proves the above claim.

Step 2: Let X be any space satisfying (xii). We claim that if $E \subset X$ is an arbitrary n-dimensional subspace and if $\delta_0 = (2Cd_E)^{-1}$ there is a subspace $F \subset E$ and a projection P from X onto F satisfying $\dim F \geq \delta_0 n$ and $\|P\| \leq 1/\delta_0$. Indeed, if $i : E \to X$ is the inclusion map, let $\widetilde{i} : X \to X$ be an extension such that $N(\widetilde{i}) \leq N(i)$. We have by (xii)

$$n \leq C\ N(\widetilde{i}) \leq C\ N(i),$$

hence obviously $n \leq C\ \gamma_2^*(i)d_E$, so that $\gamma_2^*(i) \geq 2\ \delta_0 n$. Therefore, we can obtain Step 2 by applying Sublemma 12.7 below. This completes Step 2.

Now let E be an arbitrary n-dimensional subspace of X. By Milman's *quotient of subspace theorem* (cf. Theorem 8.4) there are subspaces $E_2 \subset E_1 \subset E$ such that $\dim E_1/E_2 \geq n/2$ and $d_{E_1/E_2} \leq K$, when K is a numerical constant.

By Step 1, the space E_1^* satisfies (xii). Moreover, $E_1^* \supset E_2^\perp = (E_1/E_2)^*$ and $d_{E_2^\perp} \leq K$. Note that $\dim E_2^\perp \geq n/2$. Let $\theta = (2CK)^{-1}$.

By Step 2, there is a subspace $F \subset E_2^\perp$ with $\dim F \geq \theta n/2$ and a projection P from E_1^* onto F with $\|P\| \leq \theta^{-1}$. Clearly, $d_F \leq d_{E_1/E_2} \leq K$. By duality this implies that E_1 has a subspace $G = P^*(F^*)$ with $\dim G \geq \theta n/2$ and $d_G \leq K\|P\| \leq K\theta^{-1}$.

We have $X \supset E \supset E_1 \supset G$. By Step 2 again (this time applied to $G \subset X$), there is a subspace $F_1 \subset G$ and a projection P_1 from X onto F_1 such that $\dim F_1 \geq n(2Cd_G)^{-1} \geq n\theta(2CK)^{-1}$ and $\|P\| \leq 2Cd_G \leq 2\,CK\theta^{-1}$. Note that $d_{F_1} \leq d_G \leq K\theta^{-1}$. This shows that (ii) holds with $\delta_0 = \theta(2CK)^{-1}$ and $\tilde{C}_{\delta_0} = 2CK\theta^{-1}$, and this completes the proof that (xii) implies property (ii) in Theorem 12.2. ∎

Sublemma 12.7. *Let $i : E \to X$ be the inclusion map into X of an n-dimensional subspace of X. Assume $\gamma_2^*(i) > \alpha n$ for some $\alpha > 0$. Then there is a subspace $F \subset E$ and a projection $P : X \to F$ such that $\gamma_2(P) \leq 2/\alpha$ and $\dim F = [\alpha n/2]$.*

Proof: This is a routine argument. By (12.13) and the comments after it, $\gamma_2^*(i) > \alpha n$ iff there is an operator $S : X \to E$ such that $\gamma_2(S) < 1$ and $\mathrm{tr}\; Si > \alpha n$. We can find a decomposition $S = S_1 S_2$ with $S_2 : X \to \ell_2^m$ and $S_1 : \ell_2^m \to E$ such that $\|S_1\| < 1$ and $\|S_2\| < 1$. Let $R = S_{2|E}S_1$. Then $R : \ell_2^m \to \ell_2^m$ satisfies

$$\mathrm{tr}\; R = \mathrm{tr}\; S_1 S_{2|E} = \mathrm{tr}\; Si > \alpha n.$$

Let $R = U|R|$ be the polar decomposition of R. We have $\mathrm{tr}|R| \geq |\mathrm{tr}\; R| > \alpha n$ and $\|R\| \leq 1$. Replacing S_2 by U^*S_2, it is no loss of generality to assume that $R = |R|$. We may as well assume (for simplicity) $|R|$ diagonal with coefficients $1 \geq \mu_1 \geq \ldots \geq \mu_n \geq 0$.

Let $k = [\alpha n/2]$. Since $\sum \mu_i > \alpha n$, we have $\mu_k > \alpha/2$. Let $G = [e_1, \ldots, e_k] \subset \ell_2^m$ and let P_G be the orthogonal projection onto G. Consider the operator $w = (R_{|G})^{-1}P_G$. Note that $\|w\| \leq 2/\alpha$ since $\mu_i > \alpha/2$ for $i \leq k$. It is then easy to check that $P = S_1 w S_2$ is a projection from X onto a subspace $F = S_1(G) \subset E$ with $\dim F = k$ and $\gamma_2(P) \leq \|S_1\|\,\|w\|\,\|S_2\| < 2/\alpha$. This completes the proof. ∎

Corollary 12.8. *For any weak Hilbert space X, there is a constant K such that, for all $n > 1$, every n-dimensional subspace $E \subset X$ satisfies $d_E \leq K \mathrm{Log}\, n$. Actually, there is a projection $P : X \to E$ with $\gamma_2(P) \leq K \mathrm{Log}\, n$.*

Proof: Let E be as in the statement. Let $i : E \to X$ be the inclusion map. We claim that for some constant K we have for all $T : E \to E$

$$|tr\ T| \leq K \operatorname{Log} n(\gamma_2^*(iT)).$$

Indeed, assume $\gamma_2^*(iT) < 1$; then there is an operator $\widetilde{T} : X \to X$ extending iT and such that $\gamma_2^*(\widetilde{T}) < 1$. By Theorem 12.6, for some constant C we have

$$|\lambda_k(\widetilde{T})| \leq Ck^{-1};$$

hence,

$$|tr\ T| = \left|\sum_1^n \lambda_k(T)\right| \leq \sum_1^n |\lambda_k(T)| \leq \sum_1^n |\lambda_k(\widetilde{T}))| \leq C \sum_1^n k^{-1}.$$

By homogeneity, this proves the above claim. Then by the Hahn–Banach Theorem, this claim implies (cf. (12.13) and the comments following it) that there is an operator $P : X \to E$ such that $P_{|E} = I_E$ and $\gamma_2(P) \leq K \operatorname{Log} n$.

Clearly, P is a projection onto E and $d_E \leq \gamma_2(P) \leq K \operatorname{Log} n$. ∎

Remark: It would be interesting to know if the estimate of the preceding corollary can be improved asymptotically. For instance, any $o(\operatorname{Log}\ n)$ upper bound in the preceding statement would imply the reflexivity of X (cf. [P4], p. 348), which is proved in Chapter 14 by a rather indirect route. In the known examples related to the Tsirelson space (cf. the subsequent chapter) a better estimate can be obtained. More generally, if E possesses a λ-unconditional basis the estimate of Corollary 12.8 can be considerably improved using Proposition 12.6 and the fact that all the block sequences satisfy (uniformly) the second property in Theorem 12.3. One then gets a bound of the following form (for arbitrary $m > 1$) $d_E \leq \phi_m(\lambda) \operatorname{Log}_m n$ where $\operatorname{Log}_m n = \operatorname{Log} \operatorname{Log} \operatorname{Log} \ldots \operatorname{Log} n$ (m times) and where $\phi_m(\lambda)$ is a function depending on λ, m and X. We suspect that there exist weak Hilbert spaces without an unconditional basis and for which the estimate of Corollary 12.8 cannot be improved to a function which is (say) $o(\operatorname{Log} \operatorname{Log} n)$.

Remark 12.9: In a Hilbert space H, we have for $x_1, \ldots, x_n, x_1^*, \ldots, x_n^*$ in the unit ball of H

$$(12.21) \qquad |\det(< x_i^*, x_j >)|^{1/n} \leq 1.$$

Indeed, if M is the matrix $(< x_i^*, x_j >)$, we have

$$|\det(M)|^{1/n} = |\det |M||^{1/n} = |\prod_1^n \lambda_k(|M|)|^{1/n},$$

hence

$$\leq \frac{1}{n} \sum \lambda_k(|M|) = \frac{1}{n}\|M\|_{C_1}$$

$$\leq \frac{1}{n} \left(\sum \|x_i\|^2 \sum \|x_j^*\|^2\right)^{1/2} \leq 1.$$

Conversely, it can be shown that if (12.21) holds for $n = 2$ and if $\dim X \geq 3$ then X is isometric to a Hilbert space. This is proved in [Di], cf. also [Am]. This shows that there does not seem to be any candidate for an isometric version of the notion of weak Hilbert space.

Remark: There are several alternate possible proofs of the various implications between the twelve properties which we proved in this chapter all to be equivalent. For instance, there are direct simple proofs for (vii) $\Rightarrow$ (xi) (this yields a different proof of (vii) $\Rightarrow$ (i)) and also for (ix) $\Rightarrow$ (vii) or (x) $\Rightarrow$ (v).

Notes and Remarks

Most of this chapter is based on [P5], although several statements which merely combine a type 2 and a cotype 2 result are already implicitly in [MP1], for instance Theorem 12.3. Thus, most of Theorem 12.2 is in [P5].

Proposition 12.4 reformulates an unpublished lemma of Gordon–Lewis (communicated to us by W. B. Johnson). Their lemma said that if an n-dimensional normed space X with a 1-unconditional basis contains a λ-isomorphic and λ-complemented copy of $\ell_2^{[\delta n]}$, then it also contains a λ'-isomorphic copy of $\ell_2^{[\delta' n]}$ which is spanned by a subset of the basis vectors (with λ' and δ' depending on λ and δ only).

Theorem 12.5 comes from [MP1] and [Pa1]. The implication (i) $\Rightarrow$ (vii) is essentially in [MP1], while the converse is due to A. Pajor [Pa1]. Theorem 12.6 and Corollary 12.8 come from [P5]. Finally, Sublemma 12.7 is a known duality argument.

Chapter 13

Some Examples: The Tsirelson Spaces

In this chapter we present an example of a weak Hilbert space which is not isomorphic to a Hilbert space, namely we will prove

Theorem 13.1. *There is a Banach space X with a 1-unconditional basis which is a weak Hilbert space (actually X is both type 2 and weak cotype 2) but is not isomorphic to ℓ_2.*

Note: Actually, we will build a family $\{X_\delta | 0 < \delta < 1\}$ of such spaces with additional properties given in the remarks at the end of this chapter.

Notation: Let X be a space with a 1-unconditional basis (e_k). Recall that this means that X is the closed span of (e_k) and that for all finitely supported scalar sequences (α_k) and (β_k) we have

$$(13.1) \qquad (|\alpha_k| \le |\beta_k| \quad \forall\, k \in \mathbf{N}) \Rightarrow \left\| \sum \alpha_k e_k \right\| \le \left\| \sum \beta_k e_k \right\|.$$

Let us denote by (e_k^*) the linear functionals on X which are biorthogonal to (e_k).

For x, y in X, we will write simply $x \le y$ whenever $e_k^*(x) \le e_k^*(y)$ for all k in $\mathbf{N}$.

Moreover, for all $x = \sum x_k e_k$ in X we denote $|x|$ the element of X defined by $|x| = \sum |x_k| e_k$. Let $0 < p \le \infty$. Let $x_1, \ldots, x_n$ be a finite sequence in X. Let $(x_{ik})_k$ be the coordinates of x_i so that $x_i = \sum x_{ik} e_k$.

We will denote by

$$\left(\sum_1^n |x_i|^p \right)^{1/p}$$

the element of X defined by

$$(13.2) \qquad \left(\sum_1^n |x_i|^p \right)^{1/p} = \sum_k \left(\sum_{i=1}^n |x_{ik}|^p \right)^{1/p} e_k,$$

with the usual convention when $p = \infty$.

205

We will use below the notion of "2-convexity" (cf. [LT2] for more details). Recall that X is called p-convex if for all x, y in X we have

$$\| (|x|^p + |y|^p)^{1/p} \| \leq (\|x\|^p + \|y\|^p)^{1/p}.$$

We will denote by P_n the canonical projection of $\mathbf{R}^{(\mathbf{N})}$ onto the span of $\{e_0, e_1, \dots, e_n\}$. More precisely,

$$\forall \, x \in \mathbf{R}^{(\mathbf{N})} \quad P_n x = (x_0, x_1, \dots, x_n, 0, 0, \dots).$$

We denote $\|x\|_\infty = \sup |x_k|$.

We will now define the space X of Theorem 13.1. Let $0 < \delta < 1$ be fixed throughout this discussion and most of the chapter. We will build a space X_δ possessing the properties listed in Theorem 13.1. As usual we denote by $\mathbf{R}^{(\mathbf{N})}$ the space of all finitely supported sequences of real numbers. We will define (by induction on n) a sequence of norms $\| \ \|_n$ on $\mathbf{R}^{(\mathbf{N})}$ which are 1-unconditional. By this, we mean that they all satisfy (13.1) when (e_k) is the canonical basis of $\mathbf{R}^{(\mathbf{N})}$. Equivalently, the canonical basis of $\mathbf{R}^{(\mathbf{N})}$ is a 1-unconditional basis of the completion of $\mathbf{R}^{(\mathbf{N})}$ relative to each of the norms $\| \ \|_n$. We will use the notation (13.2) for the canonical basis of $\mathbf{R}^{(\mathbf{N})}$ also.

Let us now define the norms $\| \ \|_n$. We start by $\|x\|_0 = \sup |x_k|$. Then, given $\| \ \|_n$, we define for all x in $\mathbf{R}^{(\mathbf{N})}$

$$\|x\|_{n+1} = \sup \left\{ \|x\|_n, \delta \left(\sum_{i \leq m} \|y_i\|_n^2 \right)^{1/2} \right\},$$

where the supremum runs over all finite sequences $y_1, \dots, y_m$ in $\mathbf{R}^{(\mathbf{N})}$ such that

$$(13.3) \qquad P_m y_1 = P_m y_2 = \dots = P_m y_m = 0$$

and

$$(13.4) \qquad \left(\sum_1^m |y_i|^2 \right)^{1/2} \leq |x|.$$

The first condition means that the supports of $y_1, \dots, y_m$ all lie inside the interval $[m + 1, \infty[$.

Throughout this chapter we will denote simply by $x \cdot y$ the pointwise product of two sequences x, y in $\mathbf{R}^{(\mathbf{N})}$, i.e.

$$(x \cdot y)_k = x_k y_k.$$

We may reformulate the preceding definition as follows:

$$(13.5) \qquad \|x\|_{n+1} = \sup\left\{ \|x\|_n, \delta\left(\sum_{i\le m} \|y_i \cdot x\|_n^2\right)^{1/2}\right\},$$

where the supremum runs over all finite sequences $y_1, \ldots, y_m$ satisfying (13.3) and

$$(13.6) \qquad \left\|\left(\sum_1^m |y_i|^2\right)^{1/2}\right\|_\infty \le 1.$$

It is very easy to check that (13.5) is equivalent to the original definition. We will call *acceptable* any finite set $(y_1, \ldots, y_m)$ satisfying (13.3) and (13.6).

Clearly, we have $\|x\|_0 \le \|x\|_1 \le \cdots \le \|x\|_n \le \cdots$, and moreover

$$(13.7) \qquad \|x\|_n \le \left(\sum |x_k|^2\right)^{1/2} \quad \text{for all } x \text{ in } \mathbf{R}^{(\mathbf{N})}.$$

Also, $\| \ \|_n$ is a 1-unconditional norm on $\mathbf{R}^{(\mathbf{N})}$. All this is easy to check by induction on n (the subadditivity of $\| \ \|_{n+1}$ follows by induction from the definition (13.5)). Therefore, we can define a norm on $\mathbf{R}^{(\mathbf{N})}$ by setting

$$\|x\| = \lim_{n\to\infty} \|x\|_n.$$

Clearly, this is a 1-unconditional norm on $\mathbf{R}^{(\mathbf{N})}$ and $\|x\| \le (\sum |x_k|^2)^{1/2}$.

More generally, it can be shown that $\| \ \|_n$ and $\| \ \|$ all satisfy the "2-convexity" inequality, as follows:

$$(13.8)$$
$$\forall \ x, y \in \mathbf{R}^{(\mathbf{N})} \qquad \|(|x|^2 + |y|^2)^{1/2}\|_n \le (\|x\|_n^2 + \|y\|_n^2)^{1/2},$$
$$(13.8') \qquad \|(|x|^2 + |y|^2)^{1/2}\| \le (\|x\|^2 + \|y\|^2)^{1/2}.$$

Indeed, (13.8) is easy to show by induction on n (using (13.5)) and (13.8)$'$ follows after passing to the limit.

Let us summarize the main properties of our space.

Proposition 13.2. *For each $0 < \delta < 1$, let us denote by X_δ the completion of $\mathbf{R}^{(\mathbf{N})}$ for the norm $\| \ \|$ described above. Then (e_k) is a 1-unconditional basis of X_δ and we have*

(i) $\forall \ x \in \mathbf{R}^{(\mathbf{N})}$ $\quad \|x\|_\infty \le \|x\| \le \|x\|_{\ell_2}.$

(ii) X_δ *is 2-convex, i.e. (13.8)$'$ holds.*
(iii) *For all x in $\mathbf{R}^{(\mathbf{N})}$, we have*

$$(13.9) \qquad \|x\| = \sup\left\{\|x\|_\infty, \delta\left(\sum_{i\leq m}\|y_i\cdot x\|^2\right)^{1/2}\right\}$$

where the supremum runs over all acceptable sets $(y_1,\dots,y_m)$.

Proof: Let us prove (iii). Let us denote by $\|\|x\|\|$ the right side of
(13.9). Then by induction on n we find (since $\|\ \|_n \leq \|\ \|$) that
$\|x\|_{n+1} \leq \|\|x\|\|$ for all $n \geq 0$, hence $\|x\| \leq \|\|x\|\|$. The converse
inequality is also an immediate consequence of (13.5). ∎

To prove Theorem 13.1, we will need several lemmas. The first one
is a known result of independent interest. We recall here the notation
we use for all m-dimensional normed space E

$$d_E = d(E,\ell_2^m).$$

Lemma 13.3. *Let m be a fixed integer.*

(i) *Let Y be a Banach space. Let C be a constant such that for all
m-tuples $y_1,\dots,y_m$ in Y, and for all orthogonal $m\times m$ matrices
(a_{ij}), we have*

$$\left(\sum_{i=1}^m\left\|\sum_{j=1}^m a_{ij}y_j\right\|^2\right)^{1/2} \leq C\left(\sum\|y_j\|^2\right)^{1/2}.$$

Then for all m-dimensional subspaces $E\subset Y$, we have $d_E \leq 4C$.

(ii) *Let Y be a space with a 1-unconditional basis. Assume that
there are positive constants a and b such that, for all m-tuples
$y_1,\dots,y_m$ in Y, we have*
(13.10)
$$\frac{1}{a}\left(\sum_1^m\|y_j\|^2\right)^{1/2} \leq \left\|\left(\sum_1^m|y_j|^2\right)^{1/2}\right\| \leq b\left(\sum_1^m\|y_j\|^2\right)^{1/2}.$$

Then for all m-dimensional subspaces $E\subset Y$ we have $d_E \leq 4ab$.

Proof: Essentially, this result first appeared in [KRT]. Using more
recent information, it can be viewed merely as a combination of a result

of Kwapién characterizing d_E in terms of 2-summing operators and of a theorem of Tomczak from [TJ3]. We give some more details for the convenience of the reader. First we observe that (ii) immediately follows from (i) and the observation that we have

$$\left(\sum_i \left|\sum_j a_{ij} y_j\right|^2\right)^{1/2} = \left(\sum |y_j|^2\right)^{1/2}$$

if (a_{ij}) is an orthogonal matrix, so that (ii) implies (i) with $C \leq ab$. Therefore it is enough to prove (i).

Kwapién (cf. [K2] or [P2], Corollary 2.10, p. 27) proved that for all m-dimensional spaces E we have

$$(13.11) \qquad d_E = \sup\{\pi_2(\beta)|\beta : \ell_2^m \to E, \pi_2(\beta^*) \leq 1\}.$$

For any operator $u : X_1 \to X_2$ between Banach spaces we define

$$\pi_2^{(m)}(u) = \sup\left(\sum_1^m \|uTe_j\|^2\right)^{1/2},$$

where the supremum runs over all operators $T : \ell_2^m \to X_1$ with $\|T\| \leq 1$, and where $e_1, \ldots, e_m$ denotes here the canonical basis of $\mathbf{R}^m$.

It is proved in [TJ3] that if u is of rank at most m then

$$(13.12) \qquad \pi_2(u) \leq 2\pi_2^{(m)}(u).$$

By duality, this has the following reformulation. Let us denote simply by K the set of all operators $w : \ell_2^m \to E$ such that $\sum_1^m \|we_j\|^2 \leq 1$, and let us denote by B the set all $v : \ell_2^m \to \ell_2^m$ such that $\|v\| \leq 1$. Then a dual reformulation of Tomczak's theorem implies that

$$(13.13) \qquad \{\beta : \ell_2^m \to E | \pi_2(\beta^*) \leq 1\} \subset \text{conv}\{2wv|w \in K, v \in B\}.$$

This together with (13.11) and (13.12) immediately implies

$$d_E \leq 2 \, \sup\left\{\pi_2^{(m)}(\beta)|\pi_2(\beta^*) \leq 1\right\}$$
$$\leq 4 \, \sup\left\{\pi_2^{(m)}(wv)|w \in K, v \in B\right\}.$$

Equivalently, we have

$$d_E \leq 4 \, \sup\left\{\left(\sum \|wTe_j\|^2\right)^{1/2} \Big| w \in K, T \in B\right\}.$$

By convexity, since the orthogonal transformations are the extreme points of B, we have

$$(13.14) \quad d_E \leq 4 \, \sup \left\{ \left(\sum \|wAe_j\|^2 \right)^{1/2} \middle| w \in K, A \text{ orthogonal} \right\}.$$

Finally, letting $y_j = w(e_j)$, writing $Ae_j = \sum_{i=1}^m a_{ij} e_i$ (and using the fact that the transposed matrix (a_{ji}) is also orthogonal), we deduce from (13.14) that $d_E \leq 4C$, which completes the proof.

Remark 13.4: By a known duality argument, it is possible to show that if Y is as in (i) above (resp. as in (ii) above) then for all $E \subset Y$ with $\dim E = m$, there is a projection $P : Y \to E$ satisfying $\gamma_2(P) \leq 4C$ (resp. $\gamma_2(P) \leq 4ab$). Indeed, by (13.12) and (13.13) (with Y instead of E) we have for all $\beta : \ell_2^m \to Y, \pi_2(\beta) \leq 4C\pi_2(\beta^*)$. This implies that if $i : E \to Y$ is the inclusion mapping with $\dim E = m$, then for all $u : E \to E$ we have $N(u) \leq 4C\gamma_2^*(iu)$, hence $\mathrm{tr}\, u \leq 4C\gamma_2^*(iu)$. By the Hahn–Banach theorem, it is easy to see that this implies the existence of a projection $P : Y \to E$ with $\gamma_2(P) \leq 4C$. If Y is as in (ii) we simply use $C \leq ab$.

We can now prove

Lemma 13.5. *The space X_δ is a weak Hilbert space. Actually, it is a weak cotype 2 space of type 2.*

Proof: We first show that X_δ is a weak cotype 2 space. Let Y_m be the span of $\{e_{m+1}, e_{m+2}, \dots\}$ in X_δ. Clearly, by Proposition 13.2, Y_m satisfies the assumption of Lemma 13.3 with $a = \delta^{-1}$ and $b = 1$. Therefore, every m-dimensional subspace $E \subset Y_m$ satisfies $d_E \leq 4\delta^{-1}$.

Now, consider a subspace $E_1 \subset X_\delta$ with $\dim E_1 = 2m + 1$, and let $E_2 = E_1 \cap Y_m$. Clearly, $\dim E_2 \geq m$. Therefore E_2 (and hence E_1) contains an m-dimensional subspace E which must satisfy $d_E \leq 4\delta^{-1}$. By Theorem 10.2, this proves that X_δ is a weak cotype 2 space. Therefore, X_δ is of cotype q for all $q > 2$ (by Proposition 10.7) and since it is 2-convex, it must actually be of type 2 by a result of Maurey (cf. [LT2], 1.f.3 and 1.f.9).

We note in passing that if we use Remark 13.4 and the fact that Y_m is complemented in X_δ by a norm one projection we obtain that the subspace $E \subset E_2$ in the preceding argument is also $4\delta^{-1}$-complemented in X_δ. This gives an alternate proof (based on Theorem 12.1) that X_δ is a weak Hilbert space.

To prove that X_δ is not isomorphic to a Hilbert space, we will need several more specific lemmas concerning the norms $\|\ \|_n$.

Lemma 13.6. *For all x in $\mathbf{R}^{(\mathbf{N})}$ and all $n \geq 0$ we have*

(13.15)
$$\|x\|^2 \leq \|x\|_n^2 + \delta^{2n}\|x\|_{\ell_2}^2.$$

Proof: This is easy by induction on n. The case $n = 0$ is trivial. Assume (13.15) is proved for n and let $(y_i)_{i \leq m}$ be an acceptable set. We have by (13.15)

$$\delta^2 \sum \|x \cdot y_i\|^2 \leq \delta^2 \sum \|x \cdot y_i\|_n^2 + \delta^{2n+2}\sum \|x \cdot y_i\|_{\ell_2}^2$$
$$\leq \|x\|_{n+1}^2 + \delta^{2n+2}\|x\|_{\ell_2}.$$

By (13.9) this implies (13.15) for the integer $n + 1$. q.e.d.

In the next lemma we denote by $\operatorname{supp}(x)$ the support of an element x in $\mathbf{R}^{(\mathbf{N})}$, i.e.

$$\operatorname{supp}(x) = \{k \,|\, x_k \neq 0\}.$$

Lemma 13.7. *Let x_0, x_1 in $\mathbf{R}^{(\mathbf{N})}$ be such that for some integer N_0 we have $\operatorname{supp}(x_0) \subset [0, N_0]$ and $\operatorname{supp}(x_1) \subset]N_0, \infty[$. Then for all $n \geq 0$*

(13.16) $\|x_0 + x_1\|_{n+1}^2 \leq \max\{\|x_0\|_{n+1}^2 + \delta^2 N_0 \|x_1\|_n^2, \quad \|x_1\|_{n+1}^2\}.$

Proof: If $\|x_0 + x_1\|_{n+1} = \|x_0 + x_1\|_\infty$, (13.16) is clear. Otherwise, let $(y_i)_{i \leq m}$ be an acceptable set. Let $x = x_0 + x_1$. We wish to majorize $\delta^2 \sum_{i \leq m} \|y_i \cdot (x_0 + x_1)\|_n^2$. There are two cases. Either $m \geq N_0$, then (13.3) implies $y_i \cdot (x_0 + x_1) = y_i \cdot x_1$, and hence

$$\delta^2 \sum \|y_i \cdot (x_0 + x_1)\|_n^2 \leq \|x_1\|_{n+1}^2.$$

Or $m < N_0$. In that case, using the 2-convexity of $\|\ \|_n$ (cf. (13.8)), we have

$$\delta^2 \sum \|y_i \cdot (x_0 + x_1)\|_n^2 \leq \delta^2 \sum \|y_i \cdot x_0\|_n^2 + \delta^2 \sum \|y_i \cdot x_1\|_n^2;$$

hence (since $|y_i \cdot x_1| \leq |x_1|$)

$$\leq \|x_0\|_{n+1}^2 + \delta^2 N_0 \|x_1\|_n^2.$$

But it is easy to check that if $x \in \mathbf{R}^{(\mathbf{N})}$

$$\|x\|_{n+1}^2 = \sup\left\{\|x\|_\infty, \delta^2 \sum \|y_i \cdot x\|_n^2\right\}.$$

where the sup runs over all acceptable sets (y_i). Therefore, (13.16) follows immediately from the two cases. ∎

As an immediate consequence, we have

Lemma 13.8. *For $0 < \delta < 1$ and $n \geq 0$ let us denote by $X_{\delta n}$ the completion of $\mathbf{R}^{(\mathbf{N})}$ for the norm $\| \; \|_n$. Then every infinite-dimensional subspace of $X_{\delta n}$ contains a subspace isomorphic to c_0. In particular, $X_{\delta n}$ has no subspace isomorphic to ℓ_2.*

Proof: We prove this by induction on n. The case $n = 0$ is a well-known property of the space c_0 (cf. [LT1], Prop. 2.a.2). Assume Lemma 13.8 known for n and let us prove it for $n + 1$. Let Y be a subspace of $X_{\delta n+1}$. We may assume without loss of generality that Y is spanned by a sequence of consecutive blocks of the basis (e_k) (cf. [LT1], 1.a.11). If $\| \; \|_{n+1}$ is equivalent to $\| \; \|_n$ on Y we have nothing to prove. Otherwise, it follows clearly from Lemma 13.7 that, for each $\varepsilon > 0$, we can find a sequence $x_0, x_1, \ldots$ in $\mathbf{R}^{(\mathbf{N})}$ of disjoint consecutive blocks inside Y such that $\|x_k\|_{n+1} = 1$ for all k and

$$\|x_0 + \ldots + x_k\|_{n+1} < 1 + \varepsilon.$$

(Indeed, given $x_0, \ldots, x_k$ with support in $[0, \ldots, N_0]$ and satisfying this, we can find x_{k+1} in Y with support in $]N_0, \infty[$ such that $\|x_{k+1}\|_{n+1} = 1$ and with $\|x_{k+1}\|_n$ small enough so that $\|x_0 + \ldots + x_k\|_{n+1}^2 + \delta^2 N_0 \|x_{k+1}\|_n^2 < 1 + \varepsilon$.) The sequence (x_k) constructed in this way obviously spans a subspace of Y which is $(1 + \varepsilon)$-isomorphic to c_0. This proves the announced result for $n + 1$. ∎

Proof of Theorem 13.1: By Lemma 13.5, it remains only to prove that X_δ is not isomorphic to ℓ_2. Assume to the contrary that X_δ is isomorphic to ℓ_2. Then (since (e_k) is an unconditional basis of X_δ) there is necessarily a number $\theta > 0$ such that

$$\forall \, x \in \mathbf{R}^{(\mathbf{N})} \qquad \theta \left(\sum |x_k|^2 \right)^{1/2} \leq \|x\|.$$

By Lemma 13.6, this implies

$$\theta^2 \sum |x_k|^2 \leq \|x\|_n^2 + \delta^{2n} \sum |x_k|^2.$$

Choosing n large enough so that $\delta^{2n} < \theta^2/2$, we obtain

$$\forall \, x \in \mathbf{R}^{(\mathbf{N})} \qquad \left(\frac{\theta^2}{2} \right) \sum |x_k|^2 \leq \|x\|_n^2,$$

but this and (13.7) implies that $X_{\delta n}$ is isomorphic to ℓ_2, which contradicts Lemma 13.8. This contradiction concludes the proof. ∎

Remarks:

(i) The preceding argument shows that no subsequence of (e_k) is equivalent to ℓ_2. Actually, it is known (cf. [Jo2]) that X_δ *contains no subspace isomorphic to ℓ_2.* (The relation between X_δ and Johnson's space is discussed in detail below.)

(ii) In fact, it is proved in [CJT] that any sequence of normalized disjoint *consecutive* blocks on the basis (e_k) is equivalent to a subsequence of (e_k) in X_δ (and hence cannot span an isomorph of ℓ_2).

(iii) Morover, it is known (cf. [C]) that the spaces X_δ are all mutually non-isomorphic, and in fact are *totally incomparable*, i.e. if $\delta \neq \delta^{'}$, then X_δ and $X_{\delta'}$ have no isomorphic infinite-dimensional subspaces.

Remark 13.9: The spaces X_δ described above are part of a class of examples known as the "Tsirelson spaces". They are named after the Soviet mathematician Boris Tsirelson (sometimes spelt also Čirelson) who constructed the first example of a Banach space which contains no isomorphic copy of c_0 or $\ell_p (1 \leq p < \infty)$, cf. [T]. (See also [FJ2] or [LT1], 2.e.1). Many variants of his construction have been investigated and a lot of information is available concerning these spaces (cf. [CS]).

In particular, several equivalent definitions of the norm are known. Let us say that a subset $(y_1, \ldots, y_m)$ is *allowable* if there are disjoint subsets $E_1, \ldots, E_m$ of $\mathbf{N}$ such that y_i coincides with the indicator function of E_i and such that

$$E_1 \cup E_2 \cup \ldots \cup E_m \subset [m+1, \infty[.$$

We will prove below the following

Claim: If we repeat the preceding construction with *allowable* subsets instead of *acceptable* ones, we obtain the same norms $\| \ \|_n$ and $\| \ \|$ and the same space X_δ.

Indeed, let us denote by $[\ ]_n$ the sequence of norms constructed as above but with allowable subsets of $\mathbf{R}^{(\mathbf{N})}$. We set $[x] = \lim_{n \to \infty} [x]_n$ as before. Clearly, $[\ ]$ still satisfies the 2-convexity property (13.8). Hence, if we define

$$(13.17) \qquad \forall \, x \in \mathbf{R}^{(\mathbf{N})} \qquad |||x||| = \left[\sum |x_k|^{1/2} e_k \right]^2 ,$$

we obtain a 1-unconditional *norm* on $\mathbf{R}^{(\mathbf{N})}$. By the defining property of [ ], we have

$$(13.18) \qquad \forall\, x \in \mathbf{R}^{(\mathbf{N})} \qquad \delta^2 \sum_{i \leq m} |||y_i \cdot x||| \leq |||x|||$$

for all $y_1, \ldots, y_m$ allowable.

Now if we fix m and an integer N, we may consider the convex hull of all the allowable m-tuples $(y_1, \ldots, y_m)$ with $\mathrm{supp}(y_i) \subset [m + 1, m + N]$. It is easy to check that this convex hull coincides with the set C of all the m-tuples $(z_1, \ldots, z_m)$ in $\mathbf{R}^{(\mathbf{N})}$ such that $\mathrm{supp}(z_i) \subset [m + 1, m + N]$, $z_i \geq 0$ and $\|\sum_{i \leq m} z_i\|_\infty \leq 1$. (Indeed, the allowable sets are the extreme points of C.)

Therefore, by convexity we deduce from (13.18) that for all $z_1, \ldots, z_m$ in C

$$(13.19) \qquad \delta^2 \sum_{i \leq m} |||z_i \cdot x||| \leq |||x|||.$$

Let us denote for all x in $\mathbf{R}^{(\mathbf{N})}_+$, $x^{1/2} = \sum x_k^{1/2} e_k$. Going back to $[x]$, (13.19) implies

$$(13.20) \qquad \delta \left(\sum \left[z_i^{1/2} \cdot x \right]^2 \right)^{1/2} \leq [x].$$

But clearly, when $(z_1, \ldots, z_m)$ runs over C, then $(z_1^{1/2}, \ldots, z_m^{1/2})$ runs over all *acceptable* subsets with support in $[m + 1, m + N]$.

Therefore, since N is arbitrary, (13.20) implies

$$\delta \left(\sum [y_i \cdot x]^2 \right)^{1/2} \leq [x]$$

for all $(y_1, \ldots, y_m)$ *acceptable*.

This clearly implies $\|x\| \leq [x]$, but the converse implication is trivial (since allowable $\Rightarrow$ acceptable) so that we conclude as announced that $[x] = \|x\|$. Similarly, we have $[x]_n = \|x\|_n$ for each n. This proves the above claim.

We can restrict the notion of allowability further and still obtain the same space. More precisely, let us say that $(y_1, \ldots, y_m)$ is an *admissible* subset of $\mathbf{R}^{(\mathbf{N})}$ if it is allowable and if the sets $E_1, \ldots, E_m$ which are the supports of $y_1, \ldots, y_m$ satisfy

$$\max\{k \in E_i\} < \min\{k \in E_{i+1}\}$$

for all i.

Equivalently, this means that we restrict $y_1, \ldots, y_m$ to be the indicator functions of *consecutive* disjoint subintervals of $[m + 1, \infty[$.

We can then repeat the original construction of X_δ with admissible sets instead of acceptable or allowable ones, and we thus obtain a new norm on $\mathbf{R}^{(\mathbf{N})}$. Casazza and Odell proved ([CO]) the surprising fact that this new norm is actually equivalent to the norm of X_δ as defined above with allowable (or acceptable) sets.

Notes and Remarks

In the abundant literature on Tsirelson's spaces (see [CS]) the completion of $\mathbf{R}^{(\mathbf{N})}$ for the norm defined in (13.17) with $\delta = 2^{-1/2}$ is traditionally denoted by T and is called the Tsirelson space. Tsirelson's original example is the dual space T^*. The space X_δ with $\delta = 2^{-1/2}$ is denoted by T_c and is called the 2-convexified version of T (or rather of the *modified* version of T) in the book [CS].

W. B. Johnson ([Jo2]) proved that this space T_c has the properties required in Theorem 13.1. Many variations are possible. In particular, the consideration of subsequences of the basis yields examples with significant strengthenings of the weak cotype 2 property. Namely, the *proportional dimension* δn of the Euclidean subspaces of an n-dimensional subspace can be replaced by $n - f(n)$ for any function f tending to infinity with n. See [FLM] for more details on this.

The reader will also find in [Jo2] a proof that X_δ does not contain any isomorphic copy of ℓ_2. (Actually, no quotient of a subspace of X_δ is isomorphic to ℓ_2; cf. [Jo2], [Jo3].) If one uses the *admissible* version of the space X_δ, a proof that X_δ does not contain ℓ_2 was given already in [FJ2] but it was realized only later in [CO] that this modification leads to the same Banach spaces.

Many results on block basic sequences in these spaces can be found in [CJT] and [BCLT]. In particular, it is proved in [CJT] that T^* is *minimal*, i.e. it embeds in every infinite-dimensional subspace of itself. In [BCLT], it is proved that T has a unique unconditional basis (up to permutation).

The subsequences of the basis of X_δ which are equivalent to the original basis are characterized in [B2]; see also [B1]. The Tsirelson space is also used in Tzafriri's construction of a Banach space with equal norm cotype q which is not of cotype q ($q > 2$); see [Tz].

It is proved in [Jo3] that there is a subspace Y of X_δ such that all of its subspaces have a basis. More precisely, Y is spanned by a subsequence (e_{n_k}) of the basis of X_δ with (n_k) growing sufficiently fast so that every quotient of a subspace of Y has a basis (but is not isomorphic to ℓ_2). See [Jo3] for the details. (Actually, by the results of [CO] this subspace Y is isomorphic to X_δ; see [CS] for more precision.)

Theorem 13.1 and Proposition 13.2 are due to Johnson [Jo2]. We have modified the presentation to make it as brief and as transparent as possible, but the ideas (in particular, Lemmas 13.5, 13.6, 13.7 and their proofs) are all explicitly or implicitly either in [Jo2] or in [FJ2]. Finally, the proof of Lemma 13.3 is a combination of well-known results of Kwapién [K2] and an inequality on 2-summing operators of rank m due to Tomczak-Jaegermann [TJ3], but the result first appeared (essentially) in [KRT].

Chapter 14

Reflexivity of Weak Hilbert Spaces

In this chapter we discuss the proof and some ideas related to the following result of W. B. Johnson (we incorporate an observation of Bourgain which lifted an unnecessary restriction).

Theorem 14.1. ([Jo1]) *Weak Hilbert spaces are reflexive.*

Since being a weak Hilbert space is clearly a superproperty in the sense of James [J1], this implies that weak Hilbert spaces are super-reflexive, hence by [E] have an equivalent uniformly convex and uniformly smooth norm. We know of no satisfactory estimates for the corresponding moduli of convexity or smoothness (see [P4] for related information).

The proof of Johnson's result leads naturally to at least two (*a priori* different) weakenings of the notion of Hilbert space.

We first recall that (for $\lambda \geq 1$) a finite or infinite sequence (x_i) in a Banach space is called a λ-unconditional basic sequence if for all finitely supported sequences of scalars (α_i) and for all choices of signs $\varepsilon_i = \pm 1$ we have

$$(14.1) \qquad \left\| \sum \varepsilon_i \alpha_i x_i \right\| \leq \lambda \left\| \sum \alpha_i x_i \right\|.$$

Note that if $\|x_i\| \geq 1$ this implies

$$(14.1)' \qquad \sup |\alpha_i| \leq \lambda \left\| \sum \alpha_i x_i \right\|.$$

We will say that a Banach space X possesses the property (H) if for each $\lambda > 1$ there is a constant $K(\lambda)$ such that for any n and any λ-unconditional basic sequence $(x_1, \dots, x_n)$ with $\|x_i\| = 1 (i = 1, \dots, n)$ we have

$$(14.2) \qquad K(\lambda)^{-1} n^{1/2} \leq \left\| \sum_{1}^{n} x_i \right\| \leq K(\lambda) n^{1/2}.$$

Let $\alpha = (\alpha_n)$ be a sequence of numbers tending to 0 and let (α_n^*) be the non-increasing rearrangement of $\{|\alpha_n| \mid n \geq 1\}$. Let $\|\alpha\|_{2\infty} =$

$\sup n^{1/2}\alpha_n^*$ and $\|\alpha\|_{21} = \sum n^{-1/2}\alpha_n^*$. The Lorentz space $\ell_{2\infty}$ (resp. ℓ_{21}) is classically defined as the space of sequences $\alpha = (\alpha_n)$ such that $\|\alpha\|_{2\infty} < \infty$ (resp. $\|\alpha\|_{21} < \infty$).

It is easy to check that if X satisfies (H) then for all normalized λ-unconditional basic sequence $(x_1, \ldots, x_n)$ in X, we have
(14.3)

$$\forall\,(\alpha_i) \in \mathbb{R}^n \quad (\lambda K(\lambda))^{-1}\|\alpha\|_{2\infty} \leq \left\|\sum_1^n \alpha_i x_i\right\| \leq 2\lambda K(\lambda)\|\alpha\|_{21}.$$

For the proof of Theorem 14.1, we note first a simple result.

Proposition 14.2. *Weak Hilbert spaces possess the property (H).*

Proof: It follows from the second part of Therem 12.3 that weak Hilbert spaces satisfy (14.3). But (14.3) clearly implies (and is in fact equivalent to) the property (H). ∎

Note: We do not know whether conversely $(H) \Rightarrow$ weak Hilbert.

We have seen in Chapter 12 that if $\mathrm{Rad}(X)$ or $\ell_2(X)$ is weak Hilbert then X is isomorphic to a Hilbert space. The next statement shows that this can be generalized to unconditional and subsymmetric infinite sums of copies of X.

We need to recall what this means.

Let E be a Banach space and let $E^{(\mathbf{N})}$ be the space of all finitely supported sequences of elements of E. Consider on $E^{(\mathbf{N})}$ a norm $\|\ \|$ for which there are numbers $a > 0$ and $b > 0$ such that
(14.3)

$$\forall\,x = (x_0, x_2, \ldots,) \in E^{(\mathbf{N})} \quad a \sup\|x_n\| \leq \left\|\sum x_n\right\| \leq b\sum\|x_n\|.$$

The completion of $E^{(\mathbf{N})}$ for such a norm will be called here an *infinite sum of copies of E*. We denote it by Z. Moreover, if we have for all integers n_1, and all x in $E^{(\mathbf{N})}$,

$$\|x\| = \|(x_0, \ldots, x_{n_1}, 0, 0, \ldots, 0, x_{n_1+1}, x_{n_1+2}, \ldots)\|$$

(with an arbitrary finite number of zeros), then Z is called a *subsymmetric sum of copies of E*. Also, if for all x in $E^{(\mathbf{N})}$ we have

$$\forall\,(\varepsilon_n) \in \{-1, 1\}^{\mathbf{N}} \quad \|(\varepsilon_0 x_0, \ldots, \varepsilon_n x_n, \ldots,)\| \leq \lambda\|x\|,$$

we will say that Z is a *λ-unconditional sum of copies of E*.

For brevity we will say simply that Z is a U.S. sum of copies of E if it is 2-unconditional and subsymmetric.

Let Z be a subsymmetric sum of copies of E. For each integer n and for x in E, let $S_n(x) \in Z$ be defined by

$$S_n(x) = (0, 0, \dots, x, -x, 0, 0, \dots),$$

where the only non-zero coordinates are $x, -x$ placed respectively at the $2n^{\text{th}}$ and the $(2n + 1)^{\text{th}}$ place.

Then by a result of [BS] (cf. also [MS], p. 72), we have

$$\forall n \quad \forall (x_0, x_1, \dots) \in E^{\mathbf{N}} \quad \forall (\varepsilon_i) \in \{-1, +1\}^n$$

(14.4)
$$\left\| \sum_0^n \varepsilon_i S_i(x_i) \right\| \le 2 \left\| \sum_0^n S_i(x_i) \right\|.$$

Let $\tilde{E} = S_o(E) = \{(x, -x, 0, 0, \dots) | x \in E\}$ and let $\tilde{Z}$ be the closed span in Z of

$$\left\{ \sum_0^n S_i(x_i) | n \ge 1, x_i \in E \right\}.$$

By (14.4), $\tilde{Z}$ is a U.S. sum of copies of $\tilde{E}$. Note that by (14.3) we have

$$\forall x \in E \quad a\|x\| \le \|S_1(x)\| \le 2b\|x\|,$$

so that

(14.5)
$$d(E, \tilde{E}) \le 2ba^{-1}.$$

We now extend Theorem 12.3.

Theorem 14.3. ([Jo1]) *Let Z be a U.S. sum of copies of a space E. If Z satisfies the property (H) then E is isomorphic to a Hilbert space. Moreover, we have $d_E \le C$ for a constant C depending only on a, b and the constants involved in (H).*

Proof: For a sequence x in $E^{(\mathbf{N})}$, let

$$\|x\|_{\ell_{2\infty}(E)} = \|\{\|x_n\|\}\|_{2\infty} \text{ and } \|x\|_{\ell_{21}(E)} = \|\{\|x_n\|\}\|_{21}.$$

By (14.2), there is a constant C such that, for all $x = (x_1, x_2, \dots,)$ in Z,

(14.6) $$\frac{1}{C}\,\|x\|_{\ell_{2\infty}(E)} \le \|x\|_Z \le C\|x\|_{\ell_{21}(E)}.$$

By a result of Kahane (cf. e.g. [LT2], p. 74) the norms of $L_p(E)$ are all equivalent on $\mathrm{Rad}(E)$ for all $0 < p < \infty$. *A fortiori* the norms of the Lorentz spaces $L_{2\infty}(E)$ and $L_{21}(E)$ are equivalent on $\mathrm{Rad}(E)$. Note that for $x_1, \ldots, x_n$ in E we have

$$\left\|\sum_1^n \varepsilon_i x_i\right\|_{L_{2\infty}(E)} = 2^{-n/2}\left\|\left(\sum_1^n \varepsilon_i x_i\right)_{\varepsilon\in\{-1,+1\}^n}\right\|_{\ell_{2\infty}(E)},$$

and similarly for $L_{21}(E)$.

This with (14.6) implies that the space $\mathrm{Rad}_n(E)$ can be embedded into Z, uniformly with respect to n. Hence the space $\mathrm{Rad}(E)$ possesses property (H). We thus conclude by Theorem 12.3 that E is Hilbertian. The last part of the statement is clear. ∎

We need to introduce one more notion. We will say that a Banach space X is asymptotically Hilbertian (as. Hilbertian for short) if there is a constant β such that for all n there is a subspace $Y_n \subset X$ of finite codimension such that *every* n-dimensional subspace $E \subset Y_n$ satisfies $d_E \le \beta$.

It is easy to deduce from James' characterizations of reflexivity (cf. [J]) that as. Hilbertian implies reflexive; considering for instance the finite tree property of [J], one easily shows that for n large enough the space Y_n (in the above) must be reflexive, hence X itself is reflexive. Therefore, Theorem 14.1 follows from the next result due to Johnson.

Theorem 14.4.   ([Jo1]) *Every space with property (H) is asymptotically Hilbertian.*

Proof: We proceed by contradiction. Assume that there is a space X with property (H) which is not as. Hilbertian. Then for each large number β (to be specified at the end) there is in integer n such that every finite-codimensional $Y \subset X$ contains a subspace $E \subset Y$ with $\dim E = n$ and $d_E > \beta$.

Using the well-known fact that for each $\varepsilon > 0$, any f.d. subspace E is $(1 + \varepsilon)$-complemented in some finite-codimensional Y with $E \subset Y \subset X$, we can then find a sequence $\{E_k\}$ of subspace of X with $\dim E_k = n, d_{E_k} > \beta$ and generating a Schauder decomposition for its span $S = \cup_k(E_1 + \ldots + E_k)$. By compactness, we may as well assume that $d(E_i, E_j) < 2$ for all i and j. Then S is isomorphic to an infinite sum of copies of E_1 which we denote by Z. Applying the Brunel–Sucheston procedure [BS] (cf. [MS], p. 72), we can find by a suitable extraction a subsymmetric sum of copies of E_1 which is finitely representable into Z.

Let $\widetilde{E} = \{(x, -x, 0, 0, \ldots,) | x \in E_1\}$ as above.

By the result recalled after (14.4), we finally find a U.S. sum (denoted by $\widetilde{Z}$) of copies of $\widetilde{E}$ such that $\widetilde{Z}$ is isomorphic to a space finitely representable into X and hence $\widetilde{Z}$ satisfies (H). Applying Theorem 14.3 (note that a and b are here fixed numerical constants), we find $d_{\widetilde{E}} \leq C'$ for some fixed constant C'. Therefore, by (14.5), $d_E \leq C''$ for some numerical constant C'', and hence $\beta \leq C''$.

This is the desired contradiction since C'' is a constant independent of β. This concludes the proof.

Remark: *A posteriori*, it follows from Theorem 14.4 that if an arbitrary *infinite sum of copies of* X satisfies property (H) then X is isomorphic to a Hilbert space.

Remark: The converse to Theorem 14.4 is clearly false. For instance, let $X_n = \ell_{p_n}^{k_n}$ and $X = (X_1 \oplus X_2 \oplus \ldots)_{\ell_2}$. Then it is easy to choose $p_n \to 2$ and $k_n \to \infty$ so that X is asymptotically Hilbertian but fails (H).

Remark: The notion of as. Hilbertian is also a special case of a more general concept. Indeed, it is not hard to show that X is as. Hilbertian iff there is a sequence $Y_n \subset X$ of subspaces with finite codimension and a non-trivial ultrafilter $\mathcal{U}$ such that the ultraproduct $\Pi \, Y_n / \mathcal{U}$ is isomorphic to a Hilbert space. Given any property P, we thus can say that a space X has the property as. P if there are (Y_n) and $\mathcal{U}$ as above such that $\Pi \, Y_n / \mathcal{U}$ possesses the property P. We do not know whether weak cotype 2 implies as. cotype 2. (Same question with type instead of cotype.)

In the last few lines, we indicate that the natural extension of Theorem 14.1 to operators is not valid. Let $T : X \to Y$ be an operator between Banach spaces. We denote by weak-$\gamma_2(T)$ the following:
(14.7)
$$\text{weak-} \gamma_2(T) = \sup_n \sup\{\|BTA\|_{C_{1\infty}} \, \big| A : \ell_2^n \to X, B : Y \to \ell_2^n,$$
$$\pi_2(A^*) \leq 1, \ \pi_2(B) \leq 1\}.$$

Note that if we replace $C_{1\infty}$ by C_1 in (14.7), we obtain $\gamma_2(T)$ so that $\gamma_2(T) \geq$ weak $-\gamma_2(T)$.

An operator $T : X \to Y$ will be called a weak-γ_2 operator if weak-$\gamma_2(T) < \infty$.

Consider the operator $\sigma : \ell_1 \to \ell_\infty$ defined by

$$\forall\, \alpha = (\alpha_n) \in \ell_1 \qquad \sigma(\alpha) = \left(\sum_{k=1}^{n} \alpha_k \right)_{n \geq 1}.$$

Clearly, σ is not weakly compact and, in fact, σ is the *prototype* if a non-weakly compact operator (cf. [LP] Theorem 8.1). We have checked (using the fact that the main triangle projection in the sense of [KP] maps C_1 into $C_{1\infty}$) that weak-$\gamma_2(\sigma) < \infty$. This gives an example of a weak-γ_2 operator which is not weakly compact and hence does not factor through a weak Hilbert space.

Notes and Remarks

For this chapter we have included the references in the text. Most of the results are due to W. B. Johnson [Jo1], and were included (with his permission) in [P5]. See also [Jo4] for related results.

The essential content of Proposition 14.2 was observed in [MP1].

It would be interesting to find a simpler proof of the reflexivity of weak Hilbert spaces, although we feel that the various steps in Johnson's proof are of independent interest. Note also that as. Hilbertian spaces are super-reflexive (by the argument preceding Theorem 14.4) and hence by [E] have an equivalent uniformly convex (or uniformly smooth) norm, but we do not know any estimate of the uniform convexity or uniform smoothness of the renormings of a weak Hilbert space. In particular, we ask: if X is weak Hilbert, is X p-smooth and q-convex in the sense of [P4] for any $p < 2 < q$? We do know, however, that X is of type p and cotype q by Theorem 12.3 or Corollary 12.8. A possible improvement of Corollary 12.8 (to $\operatorname{Log} \operatorname{Log} n$ for instance) would prove the above conjecture.

Chapter 15

Fredholm Determinants and the
Approximation Property

We will abbreviate *approximation property* to A.P. Let X be a Banach space and let $1 \leq \lambda < \infty$. Recall that X has the A.P. (resp. λ-A.P.) if for every $\varepsilon > 0$ and every compact subset $K \subset X$ there is a finite rank operator u (resp. with $\|u\| \leq \lambda$) such that $\|u(x) - x\| < \varepsilon$ for all x in K. We say that X has the bounded (resp. metric) A.P. if it has the λ-A.P. for some $\lambda \geq 1$ (resp. for $\lambda = 1$). Finally, we say that X has the uniform A.P. (in short, U.A.P.) if there is a constant $1 \leq \lambda < \infty$ and a sequence of integers $\{k(n)|n \geq 1\}$ such that for every finite dimensional subspace $E \subset X$ there is an operator $u : X \to X$ with $\mathrm{rk}(u) \leq k(\dim E), \|u\| \leq \lambda$ and such that $u(x) = x$ for all x in E. The aim of this section is to prove

Theorem 15.1. *Every weak Hilbert space possesses the A.P.*

Corollary 15.2. *Every weak Hilbert space possesses the U.A.P.*

The proof of the corollary is immediate using a result of Heinrich [H] which says that a space X has the U.A.P. iff every ultrapower of X has the bounded A.P.

Indeed, by a well-known result of Grothendieck (cf. [LT1], p. 39) a reflexive space with the A.P. must have the metric A.P. Since weak Hilbert spaces are preserved ·by ultrapower, Theorems 15.1 and 14.1 imply that every ultrapower of a weak Hilbert space X has the metric A.P., hence by Heinrich's result X has the U.A.P. It would be interesting to know a *good* estimate of the *uniformity function* $n \to k(n)$ appearing in the definition of the U.A.P. Our proof gives no estimates at all. Since in a Hilbert space $k(n) = n$, it is conceivable that we have $k(n) \leq Cn$ for some constant C in every weak Hilbert space, but this is an open question. Conversely, such a property might imply weak Hilbert. This is also open.

The idea of the proof of Theorem 15.1 is to use *determinants*. As shown by Grothendieck ([G1]), a space X has the A.P. iff for every sequences (x_n^*) and (x_n) in X^* and X respectively, the conditions

(15.1) $$\sum_1^\infty \|x_n^*\| \, \|x_n\| < \infty$$

and

(15.2) $$\sum_1^\infty x_n^*(x)x_n = 0 \quad \forall \, x \in X$$

imply the condition

(15.3) $$\sum_1^\infty x_n^*(x_n) = 0.$$

The last condition states that the trace of $\sum_1^\infty x_n^* \otimes x_n$ is zero, where the trace is defined as usual first on $X^* \otimes X$ and then extended on the completion $X^* \widehat{\otimes} X$ by continuity. Since in weak Hilbert spaces, the determinant behaves well, it is natural to expect that the trace behaves well so that (15.1) and (15.2) imply (15.3). This is what will be shown below.

Before that, we recall that the A.P. fails in a general Banach space as shown by P. Enflo in 1972 (cf. [LT1]). More precisely, Szankowski showed that if every subspace of a space X has the A.P. then necessarily X is of type $2 - \varepsilon$ and cotype $2 + \varepsilon$ for every $\varepsilon > 0$ (cf. [LT2]). However, as shown by Johnson, this is false in general if $\varepsilon = 0$. Namely, Johnson [Jo3] showed that there is a space $\mathcal{X}$ such that every subspace (even every quotient of a subspace) of $\mathcal{X}$ has the A.P. (even a basis), but $\mathcal{X}$ is not isomorphic to a Hilbert space.

We note, however, that there is another example of Johnson with the same property but which is not a weak Hilbert space. Indeed the space described in [LT2], p. 112, is an ℓ_2 sum of spaces $\ell_{p_n}^{k_n}$ with $p_n > 2, p_n \to 2$ and k_n integers such that $(k_n)^{1/p_n - 1/2} \to \infty$ when $n \to \infty$. This space clearly fails property (H) of Chapter 14 and hence is not a weak Hilbert space.

To explain the proof of Theorem 15.1, we need more notation, following [G2].

Let X be a Banach space. Let $k_n(X) = \sup\{\det | < x_i^*, x_j > |\}$ where the supremum runs over all n-tuples (x_i^*) and (x_j) in the unit ball respectively of X^* and X.

We have seen above that if X is a weak Hilbert space then there is a constant C such that $k_n(X) \le C^n$ for all $n \ge 1$ (cf. Theorem

12.6). For a Hilbert space, it is easy to see that this holds with $C = 1$ (see Remark 12.9). Now assume that X is a general Banach space. Grothendieck ([G2]) and many other authors (cf. e.g. [Si1], [Si2], [Ru]) studied a notion of determinant on the projective tensor product $X^* \widehat{\otimes} X$ in order to generalize the Fredholm theory to Banach spaces.

In the sequel we follow [G2] rather closely. We only consider *complex* Banach spaces from now on, unless otherwise specified. First we note that since for every n-dimensional space E we have $d(E, \ell_2^n) \leq \sqrt{n}$, it follows that $k_n(E) \leq n^{n/2}$, and therefore for any Banach space X we have

$$k_n(X) \leq n^{n/2}.$$

Let $T : X \to X$ be a finite rank operator with eigenvalues $\lambda_1(T), \lambda_2(T)$, etc. By convention we set $\lambda_n(T) = 0$ if $n > \operatorname{rank}(T)$. Then we can obviously define $\det(1 + T)$ as follows:

$$(15.4) \qquad \det(1 + T) = \prod_{n=1}^{\infty} (1 + \lambda_n(T)).$$

Clearly this coincides with the usual notion of determinant if X is finite dimensional and it has all the usual properties of the determinant. Note that if E is any f.d. space containing the range of T and if $T_E : E \to E$ denotes the restriction of T to E then T and T_E have the same non-zero eigenvalues so that

$$\det(1 + T) = \det(1 + T_E).$$

From this identity we recover the usual properties of the determinant, in particular since for all $T, S : X \to X$ we have $(1 + T)(1 + S) = 1 + T + S + TS$, we find (for T, S of finite rank)

$$(15.5) \qquad \det(1 + T) \det(1 + S) = \det(1 + T + S + TS).$$

We will be interested in extending the function $T \to \det(1 + T)$ to the projective tensor product $X^* \widehat{\otimes} X$.

Recall that for any T in $X^* \otimes X$ the projective norm $\|T\|$ is defined as

$$\|T\|_\wedge = \inf \left\{ \sum_1^u \|x_i^*\| \, \|x_i\| \; \Big| \; T = \sum_1^n x_i^* \otimes x_i \right\}.$$

The space $X^* \widehat{\otimes} X$ is defined as the completion of $X^* \otimes X$ with respect to the projective norm $\| \; \|_\wedge$. The projective norm $\| \; \|_\wedge$ then uniquely extends to the completed tensor product $X^* \widehat{\otimes} X$ and it is easy to check that any T in $X^* \widehat{\otimes} X$ can be represented as

$$T = \sum_{n=1}^{\infty} x_n^* \otimes x_n \quad \text{with} \quad \sum_{n=1}^{\infty} \|x_n^*\| \, \|x_n\| < \infty$$

and

$$\|T\|_\wedge = \inf \left\{ \sum_{1}^{\infty} \|x_n^*\| \, \|x_n\| \right\},$$

where the infimum runs over all such representations.

To every $T = \sum_{n=1}^{\infty} x_n^* \otimes x_n$ in $X^* \widehat{\otimes} X$ as above we may associate an operator $\widetilde{T} : X \to X$ by setting $\widetilde{T}x = \sum x_n^*(x)x_n$ for all x in X. Clearly, $\widetilde{T}$ depends only on T and not on the series representation. The operators of the form $\widetilde{T}$ for some T in $X^* \widehat{\otimes} X$ are called nuclear and the nuclear norm of an operator $S : X \to X$ is defined as

$$N(S) = \inf\{\|T\|_\wedge | \, T \in X^* \widehat{\otimes} X \quad \widetilde{T} = S\}.$$

The main problem when dealing with spaces failing the approximation property is the *non-injectivity* of the map $T \to \widetilde{T}$ from $X^* \widehat{\otimes} X$ into $B(X,X)$. Therefore, until we know that a space has the A.P. we must carefully distinguish between T and $\widetilde{T}$. (Note also that $\|T\|_\wedge$ and $N(\widetilde{T})$ in general are different.)

To extend the function $T \to \det(1+T)$ to $X^* \widehat{\otimes} X$ we first need a *different* but equivalent definition of $\det(1+T)$ for a finite rank operator T. Let $n \geq 1$ be an integer. We consider the exterior product $X \wedge \ldots \wedge X$ (n-times). For any linear operators $T_i : X \to X$ ($i = 1,\ldots,n$) we define a linear operator

$$T_1 \wedge \ldots \wedge T_n : X \wedge \ldots \wedge X \to X \wedge \ldots \wedge X$$

as follows:

(15.6)

$$(T_1 \wedge \ldots \wedge T_n)(x_1 \wedge \ldots \wedge x_n) = \frac{1}{n!} \sum_\sigma \varepsilon_\sigma T_1(x_{\sigma_1}) \wedge \ldots \wedge T_n(x_{\sigma_n}),$$

where σ runs over all permutations of $\{1,\ldots,n\}$ and ε_σ denotes the signature of σ.

Clearly, (15.6) defines $T_1 \wedge \ldots \wedge T_n$ unambiguously by linearity on the whole of $X \wedge \ldots \wedge X$. Note that $(T_1,\ldots,T_n) \to T_1 \wedge \ldots \wedge T_n$ is a *symmetric* n-linear map. When $T_1,\ldots,T_n$ are finite-rank operators, then $T_1 \wedge \ldots \wedge T_n$ is also a finite-rank operator and hence its trace makes sense. We define

$$\alpha_n(T_1, \ldots, T_n) = tr(T_1 \wedge \ldots \wedge T_n).$$

Then α_n is a symmetric n linear form on $X^* \otimes X$.

If $T : X \to X$ is a finite rank operator, then

$$\alpha_n(T, \ldots, T) = \sum_{i_1 < \ldots < i_n} \lambda_{i_1}(T)\lambda_{i_2}(T) \ldots \lambda_{i_n}(T).$$

This is indeed easy to check if the non-zero eigenvalues of T are all distinct since if $x_1, \ldots, x_n$ are eigenvectors for T relative to distinct eigenvalues $\lambda_1, \ldots, \lambda_n$, then $x_1 \wedge \ldots \wedge x_n$ is an eigenvector for $T \wedge \ldots \wedge T$ relative to the eigenvalue $\lambda_1 \lambda_2 \ldots \lambda_n$. The general case follows from this by a perturbation argument.

We thus have found a different way to write $\det(1 + T)$; we have

$$(15.7) \qquad \forall\, T \in X^* \otimes X \quad \det(1 + T) = \sum_{n=0}^{\infty} \alpha_n(T, \ldots, T)$$

with the convention $\alpha_o(T, \ldots, T) \equiv 1$.

For brevity, we denote

$$\alpha_n(T) = \alpha_n(T, T, \ldots, T).$$

We will use the following elementary fact:

(15.8) If at least k operators among $T_1, \ldots, T_n$ coincide with an operator u of rank $< k$ then $\alpha_n(T_1, \ldots, T_n) = 0$.

Indeed, we even have $T_1 \wedge \ldots \wedge T_n = 0$, since if $T_1 = T_2 = \ldots = T_k = u$ and if $E = u(X)$ then $T_1 \wedge \ldots \wedge T_n$ takes its values in

$$\underbrace{E \wedge \ldots \wedge E}_{k \text{ times}} \wedge X \wedge \ldots \wedge X,$$

and if $\dim E < k$, this reduces to $\{0\}$.

We also need the following obvious identity for $x_1, \ldots, x_n \in X$, $x_1^*, \ldots, x_n^* \in X^*$:

$$(15.9) \qquad \alpha_n(x_1^* \otimes x_1, \ldots, x_n^* \otimes x_n) = \frac{1}{n!} \det(< x_i^*, x_j >).$$

Indeed, let $T_i = x_i^* \otimes x_i$, let $\omega = x_1 \wedge \ldots \wedge x_n$; then for any $y_1, \ldots, y_n$ in X we have by (15.6)

$$(15.10) \qquad (T_1 \wedge \ldots \wedge T_n)(y_1, \ldots, y_n) = \frac{1}{n!} \det(< x_i^*, y_j >) \cdot \omega.$$

Therefore, $T_1 \wedge \ldots \wedge T_n$ has rank ≤ 1 (its image is formed of multiples of ω); hence necessarily

$$(T_1 \wedge \ldots \wedge T_n)(\omega) = tr(T_1 \wedge \ldots \wedge T_n)\omega,$$

so that (15.10) implies (15.9).

Proposition 15.3. *Let X be an arbitrary Banach space. Then $\forall \, T_1, \ldots, T_n \in X^* \otimes X$*

$$(15.11) \qquad |\alpha_n(T_1, \ldots, T_n)| \leq (n!)^{-1} k_n(X) \|T_1\|_\wedge \cdots \|T_n\|_\wedge.$$

Therefore $\forall \, T \in X^* \otimes X$,

$$(15.12) \qquad |\alpha_n(T)| \leq (n!)^{-1} k_n(X) \|T\|_\wedge^n.$$

Proof: Assume $T_1 = x_1^* \otimes x_1, \ldots, T_n = x_n^* \otimes x_n$; then by homogeneity (15.9) implies

$$(n!)|\alpha_n(T_1, \ldots, T_n)| \leq k_n(X)\|x_1^*\| \, \|x_1\| \cdots \|x_n^*\| \, \|x_n\|$$
$$\leq k_n(X)\|T_1\|_\wedge \cdots \|T_n\|_\wedge.$$

By a convexity argument and the definition of $\| \; \|_\wedge$ this remains true for arbitrary $T_1, \ldots, T_n$ in $X^* \otimes X$, thus establishing (15.11). The other inequality (15.12) follows trivially. ∎

The preceding result allows us to extend α_n by continuity to the completed tensor product $X^* \widehat{\otimes} X$. Thus, by density we have now defined $\alpha_n(T_1, \ldots, T_n)$ and $\alpha_n(T)$ for $T_1, \ldots, T_n$ and T in $X^* \widehat{\otimes} X$. Of course, this extension still satisfies (15.11) and (15.12). In particular, since $k_n(X) \leq n^{n/2}$ for any X, we have

$$\left| \sum_0^\infty \alpha_n(T) \right| \leq \sum_1^\infty |\alpha_n(T)| \leq \sum_0^\infty \frac{n^{n/2}}{n!} \|T\|_\wedge^N.$$

This shows that $\sum_0^\infty \alpha_n(T)$ converges absolutely for any T in $X^* \widehat{\otimes} X$, so that we can *define* $\det(1 + T)$ as

$$\det(1 + T) = \sum_{n=0}^\infty \alpha_n(T), \text{ for } T \in X^* \widehat{\otimes} X.$$

Let $D_T(z) = \det(1 - zT)$. Then D_T is an entire function on $\mathbf{C}$ and its zeros are exactly (with multiplicity) the inverse of the non-zero eigenvalues of the nuclear operator $\tilde{T} : X \to X$ associated to T. We refer the reader to [G2] for a detailed proof.

We now turn to the special case when $k_n(X) \le C^n$ for all $n \ge 1$.

Lemma 15.4. *Assume $k_n(X) \le C^n$ for all $n \ge 1$. Consider u in $X^* \otimes X$ with $\mathrm{rk}(u) \le k$. Then for any v in $X^* \widehat{\otimes} X$ we have*

$$(15.13) \quad \forall \, n \ge 1 \quad |\det(1 + u + v)| \le \left(\sum_{j \le k} \frac{C^j}{j!} \|u\|_\wedge^j \right) \exp(C\|v\|_\wedge).$$

Proof: Assume first $v \in X^* \otimes X$. By the binomial formula,

$$\alpha_n(u + v) = \sum_{j \le n} \binom{n}{j} \alpha_n(\underbrace{u, \dots, u}_{j \text{ times}}, v, \dots, v);$$

hence by (15.8)

$$= \sum_{j \le n \wedge k} \binom{n}{j} \alpha_n(u, \dots, u, \, v, \dots v);$$

therefore by (15.11),

$$|\alpha_n(u + v)| \le \sum_{j \le n \wedge k} \frac{C^n}{j!(n-j)!} \|u\|_\wedge^j \, \|v\|_\wedge^{n-j},$$

so that

$$|\det(1 + u + v)| \le \sum |\alpha_n(u + v)|$$

$$\le \sum_{j \le k} \frac{C^j}{j!} \|u\|_\wedge^j \cdot \sum_{n \ge j} \frac{C^{n-j}}{(n-j)!} \|v\|_\wedge^{n-j},$$

which proves (15.13) for v in $X^* \otimes X$. By density and continuity, (15.13) remains true for all v in $X^* \widehat{\otimes} X$. (Note that actually (15.8) obviously remains valid for v in $X^* \widehat{\otimes} X$.)

As a consequence, we immediately have

Lemma 15.5. *Assume $k_n(X) \le C^n$ for all $n \ge 1$. Consider T in $X^* \widehat{\otimes} X$. Then for all $\varepsilon > 0$ there is a constant C_ε such that*

$$\forall \, z \in \mathbf{C} \quad |\det(1 - zT)| \le C_\varepsilon \exp(\varepsilon|z|).$$

Proof: We write $T = u + v$ with $\|v\|_\wedge < \varepsilon/2C$ and u of finite rank k. Then Lemma 15.4 immediately implies Lemma 15.5.

Following known ideas (cf. e.g. [Si1], [Si2]) we now make use of an essentially classical result of Phragmen–Lindelöf type.

(15.14) Let $f : \mathbf{C} \to \mathbf{C}$ be an entire function with $f(0) = 1$. Assume

(1) The zeros $\{z_n\}$ of f (counted with multiplicity) satisfy

$$\sum \frac{1}{|z_n|} < \infty.$$

(2) For each $\varepsilon > 0$, there is $C_\varepsilon > 0$ such that

$$\forall z \in \mathbf{C} \quad |f(z)| \le C_\varepsilon \exp \varepsilon |z|.$$

Then

$$f(z) = \prod_n (1 - z/z_n).$$

For a proof, see e.g. [RS]. We now deduce immediately

Theorem 15.6. *Let X be a Banach space such that for some $C \ge 1$ we have $k_n(X) \ge C^n$ for all $n \le 1$. We have then*

(a) *$|\det(1 + T)| \le \exp C\|T\|_\wedge$ for all T in $X^* \widehat{\otimes} X$.*
(b) *Let $D_T(z) = \det(1 - zT)$ and assume that the zeros $\{z_n\}$ of D_T satisfy $\sum \frac{1}{|z_n|} < \infty$. Then*

$$D_T(z) = \prod_{n=1}^{\infty} (1 - z/z_n).$$

(c) *The space X has the A.P. (so that we may identify $X^* \widehat{\otimes} X$ with the space of nuclear operators on X).*
(d) *Any nuclear operator $T : X \to X$ with absolutely summable eigenvalues $\{\lambda_n(T)\}$ must satisfy $\det(1-zT) = \Pi_1^\infty(1-z \cdot \lambda_n(T))$ for all z in $\mathbf{C}$ and $\operatorname{tr} T = \sum \lambda_n(T)$.*

Proof: Clearly (a) follows from (15.12) and (b) follows from the (15.14) applied to $f(z) = \det(1 - zT)$.

Let us show (c). For this we observe that $X^* \otimes X$ equipped with the projective norm obviously is a normed algebra for the composition of linear operators. By density we thus have a Banach algebra structure on $X^* \widehat{\otimes} X$. Moreover, (15.5) clearly implies (by density and continuity again) that for T, S in $X^* \widehat{\otimes} X$ (the product $T \cdot S$ being the one just defined) we have

$$(15.15) \qquad \det(1 + T + S + T \cdot S) = \det(1 + T)\det(1 + S).$$

Now consider $T = \sum_{n=1}^{\infty} x_n^* \otimes x_n$ in $X^* \widehat{\otimes} X$ with $\sum_1^{\infty} \|x_n^*\| \, \|x_n\| < \infty$ such that the associated operator $\widetilde{T} : X \to X$ is zero, i.e. we assume

$$(15.16) \qquad \forall \, x \in X \quad \widetilde{T}(x) = \sum_{n=1}^{\infty} x_n^*(x) x_n = 0.$$

Clearly, (15.16) implies that in the Banach algebra $X^* \widehat{\otimes} X$ we have

$$T \cdot (x_n^* \otimes x_n) = 0 \text{ for all } n,$$

hence

$$T \cdot T = 0.$$

Therefore we have for all z in $\mathbf{C}$ by (15.15)

$$\det(1 - zT)\det(1 + zT) = \det(1 + 0) = 1.$$

This shows that $z \to \det(1 - zT)$ has *no* zeros, hence by (15.14) $\det(1 - zT) \equiv 1$, so that $\alpha_n(T) = 0$ for all $n \geq 1$ and in particular

$$tr \; T = \alpha_1(T) = 0.$$

This proves (c). The map $T \to \widetilde{T}$ is therefore injective (indeed, $\widetilde{T} = 0$ implies $S\widetilde{T} = 0$ for all bounded $S : X \to X$, hence $tr \; ST = 0$ for all S, which implies $T = 0$). To prove (d) we use the fact that the zeros of $\det(1 - zT)$ are exactly (with multiplicity) the inverse of the non-zero eigenvalues of $\widetilde{T}$. Therefore, if $\sum |\lambda_n(\widetilde{T})| < \infty$, (15.16) implies

$$\det(1 - zT) = \prod(1 - z\lambda_n(\widetilde{T})),$$

and in particular identifying the coefficients we have $tr \; T = \sum \lambda_n(\widetilde{T})$, and also for all $n > 1$

$$\alpha_n(T) = \sum_{i_1 < \dots < i_n} \lambda_{i_1}(\widetilde{T}) \dots \lambda_{i_n}(\widetilde{T}). \quad \blacksquare$$

Remark: Part (d) is a generalization of a classical formula due to Lidskii [Li] (but apparently known to Grothendieck, see [G3], pp. 91–107), which says that $tr\, T = \sum \lambda_n(T)$ for any nuclear operator T on a Hilbert space.

Remark: In general, the eigenvalues of a nuclear operator on a Banach space X are not absolutely summable, unless (cf. [JKMR]) X is isomorphic to a Hilbert space. Some assumption of summability of $\{\lambda_n(T)\}$ is thus necessary in the above statement if we want the sum $\sum \lambda_n(T)$ to make sense. We should also point out that (by a result of Szankowski, cf. [LT2], p. 107 and [LT1], p. 35) for any $p \neq 2$ there is a nuclear operator T on ℓ_p such that $T^2 = 0$, hence $\lambda_n(T) = 0$ for all n, but with non-zero trace. This shows that part (d) above fails in ℓ_p $p \neq 2$. More generally, if X has a subspace Y which fails the A.P. then X cannot have property (d) above. Indeed, let $T \in Y^* \widehat{\otimes} Y$ be such that the associated nuclear operator $\widetilde{T} : Y \to Y$ is zero but $tr\, T \neq 0$. We have $T = \sum_0^\infty y_n^* \otimes y_n$ with $\sum \|y_n^*\|\, \|y_n\| < \infty$ and $\sum y_n^*(y_n) \neq 0$. By Hahn–Banach, we may extend y_n^* to a function x_n^* in X^* with $\|x_n^*\| = \|y_n^*\|$. Let $S = \sum x_n^* \otimes y_n \in X^* \widehat{\otimes} X$. Then clearly $S \cdot S = 0$, hence $\widetilde{S}^2 = 0$ and therefore $\lambda_n(\widetilde{S}) = 0$ for all n. However, $tr\, S = \sum x_n^*(y_n) = \sum y_n^*(y_n) \neq 0$. Thus X fails property (d) above. Actually, it is easy to check that if X satisfies (d) then every subspace of a quotient of X has the A.P.

Remark: The preceding also makes sense if we replace the identity of X by a bounded operator $u : X \to Y$. Precisely, we can define

$$k_n(u) = \sup\{\det(< y_i^*, u x_j >)\big|\quad x_j \in B_X \quad y_i^* \in B_{Y^*}\},$$

and study the operators u such that $k_n(u) \geq C^n$ for all $n \leq 1$.

We then find for all T in $Y^* \widehat{\otimes} X$

$$|\det(1 + uT)| \leq \exp C\|T\|_\wedge$$

and the same argument as above yields that if the eigenvalues of uT are summable, we have

$$tr(uT) = \sum \lambda_n(uT).$$

In particular, if $(\widetilde{uT}) = 0$, then $tr\, uT = 0$. This shows that $T \to tr\, uT$ unambiguously makes sense for any nuclear operator $T : Y \to X$ and we have $|tr(uT)| \leq N(T)$.

By a classical result (cf. [G1]), this implies that u is approximable, i.e. for all $\varepsilon > 0$ and all compact sets $K \subset X$ there is an operator $w : X \to Y$ of finite rank such that

$$\sup_K \|u(x) - w(x)\| < \varepsilon.$$

Moreover, the restriction of u to any invariant subspace is also approximable.

Finally, we can complete the proof of Theorem 12.6 in Chapter 12. We will use a classical result from the theory of entire functions.

Lemma 15.7. *Let $f : \mathbf{C} \to \mathbf{C}$ be an entire function such that $|f(0)| = 1$ satisfying for some constant C*

$$\forall\, z \in \mathbf{C} \quad |f(z)| \leq \exp C|z|.$$

Let $\{z_n\}$ be zeros of f arranged so that $0 < |z_1| \leq |z_2| \leq \ldots |z_n|$. Then we have $n/|z_n| \leq C(1 + e)$.

Proof: We can write

$$f(z) = \prod_1^n (1 - z/z_i) \cdot g(z),$$

where g is an entire function such that $g(0) = 1$. By subharmonicity for any $r > 0$ there is a z in $\mathbf{C}$ with $|z| = r$ such that $|g(z)| \geq 1$. Therefore,

$$\inf_{|z|=r} \left| \prod_1^n (1 - z/z_i) \right| \leq \exp(Cr).$$

Choosing $r = (1+e)|z_n|$, we find if $|z| = r$ $\quad |1 - z/z_i| \geq |z|/|z_i| - 1 \geq e$ hence $e^n \leq \exp(C|z_n|(1 + e))$ and therefore $n \leq C|z_n|(1 + e)$. ■

We now conclude and recapitulate

Theorem 15.8. *The following properties of a Banach space are equivalent:*

  (i) *X is a weak Hilbert space.*
 (ii) *There is a constant C such that $k_n(X) \leq C^n$ for all $n \geq 1$.*
(iii) *There is a constant C such that*

$$\forall\, n\; \forall\, T_1, \ldots, T_n \in X^* \otimes X \quad |\alpha_n(T_1, \ldots, T_n)| \leq C^n \|T_1\|_\wedge \cdots$$
$$\|T_n\|_\wedge.$$

(iv) *There is a constant C such that for all T in $X^* \widehat{\otimes} X$*

$$|\alpha_n(T)| \leq (n!)^{-1} C^n \|T\|_\wedge^n.$$

(v) *There is a constant C such that for all T in $X^* \widehat{\otimes} X$*

$$|\det(1 + T)| \leq \exp(C\|T\|_\wedge).$$

(vi) *There is a constant C such that all nuclear operators $T : X \to X$ satisfy*

$$\sup n|(\lambda_n(T))| \leq C N(T).$$

Proof: (i) $\Rightarrow$ (ii) was proved in Chapter 12 (Theorem 12.6).
(ii) $\Rightarrow$ (iii) $\Rightarrow$ (iv) follows from Proposition 15.3.
(iv) $\Rightarrow$ (v) is obvious.

Finally, (v) $\Rightarrow$ (vi) follows from the preceding lemma. Indeed, (v) implies $|\det(1 - zT)| \leq \exp(C|z| \|T\|_\wedge)$ for all T in $X^* \widehat{\otimes} X$. Let $\widetilde{T} :$ $X \to X$ be the associated operator. Since the zeros of $\det(1 - zT)$ are the inverses of the eigenvalues of $\widetilde{T}$ we have by Lemma 15.7 (assuming (v))

$$\sup n|\lambda_n(\widetilde{T})| \leq C(1 + e)\|T\|_\wedge$$

and this implies (vi).

Lastly, the implication (vi) $\Rightarrow$ (i) has already been proved (see Theorem 12.6).

Remark: Although we assumed that X is a *complex* Banach space, all the results which make sense in the *real* case are easily extended in that case also. For instance, Theorem 15.1 and Corollary 15.2 clearly hold in the real case as well.

Notes and Remarks

This chapter is based on [P5].

To a large extent, everything there boils down to the observation that the Fredholm Theory as developed by Grothendieck in [G2] works in a weak Hilbert space just as well as in a Hilbert space.

As already mentioned in the text, other references on the Fredholm determinants are [Si1] [Si2] [Ru]. Besides Grothendieck's, the names of Ruston, Smithies, Le.zański, and Sikorski are often associated with the development of the Fredholm Theory in the Banach space setting (cf. [Ru] for more precision).

Final Remarks

There are several important topics closely connected with the subject of this book which we have not included in the text. For recent developments related to Harmonic Analysis in $\mathbf{R}^n$ and the Theory of Maximal Functions, we refer the reader to [Bou1] and [Bou2].

Moreover, the following open problem has recently received a great deal of attention.

Problem: Is there a $\delta > 0$ such that for every n and every ball $B \subset \mathbf{R}^n$ with volume 1, there is a hyperplane section $H \cap B$ with

$$\mathrm{vol}_{n-1}(H \cap B) \geq \delta?$$

Equivalently, is there a constant $c > 0$ such that for all n and all balls B in $\mathbf{R}^n$ there is a hyperplane H in $\mathbf{R}^n$ for which

$$(\mathrm{vol}_{n-1}(H \cap B))^{\frac{1}{n-1}} \geq \left(1 - \frac{c}{n}\right) \mathrm{vol}_n(B)^{\frac{1}{n}}.$$

This problem is studied in particular in [Ba1], [Ba3], and [Ba5]. It also appeared in Bourgain's work [Bou1].

In general, given a ball $B \subset \mathbf{R}^n$ and $k < n$ it is interesting to compute the supremum or the infimum of

$$\mathrm{vol}_k(H \cap B),$$

when H runs over all k-dimensional subspaces of $\mathbf{R}^n$. For the cube $B_\infty = [-1,1]^n$ (i.e. the unit ball of ℓ_∞^n) very precise results are known. Indeed, in that case, Vaaler in [V] proved that for all k-dimensional subspaces $H \subset \mathbf{R}^n$

$$2^k \leq \mathrm{vol}_k(H \cap B_\infty).$$

On the other hand, K. Ball [Ba2] proved that for all hyperplanes $H \subset \mathbf{R}^n$

$$\mathrm{vol}_{n-1}(H \cap B_\infty) \leq 2^{n-1}\sqrt{2},$$

thus verifying a conjecture of Hensley [He].

These bounds are optimal. Ball [Ba4] generalized these bounds to sections of arbitrary dimension. Moreover, M. Meyer and A. Pajor extended the above results to sections of the unit ball of ℓ_p^n for $1 \leq p < \infty$. See [MeP].

Some partial results on the above problem appear in [BMMP]. See also [MPa].

In a completely different direction, we should mention the recent work of Bourgain, Lindenstrauss, and Milman ([BLM]) which contains a number of estimates closely related to entropy numbers.

Also, in [Bou3], J. Bourgain showed that there is a constant $0 < \delta < 1$ such that if all the $[\delta n]$-dimensional subspaces of an n-dimensional normed space E are mutually λ-isomorphic (i.e. we have $d(F_1, F_2) \leq \lambda$ for all $[\delta n]$-dimensional subspaces F_1, F_2 of E) then necessarily

$$d(E, \ell_2^n) \leq \phi(\lambda),$$

where $\phi(\lambda)$ is a constant depending only on λ. This interesting result has some connection with Chapter 10.

See also [BoS] and [BoT] for recent results related to the Dvoretzky–Rogers lemma and the results of Chapter 3.

Bibliography

[Al] A. D. Alexandrov. On the theory of mixed volume of convex bodies. I. *Math Sbornik* **2** (1937), 947–972. II. *Idem* (1938), 1205–1238. III. *Idem* **3** (1938), 27–66. IV. Idem **3**(1938), 227–251.

[Am] D. Amir. *Characterizations of Inner Product Spaces*. Birkhauser, 1986.

[BC] A. Badrikian and S. Chevet. *Mesures cylindriques, Espaces de Wiener et Fonctions Aléatoires Gaussiennes*. Springer Lecture Notes n° 379 (1974).

[Ba1] K. Ball. Isometric problems in ℓ_p and sections of convex sets. Ph.D. Thesis, Cambridge University, 1986.

[Ba2] ———. Cube slicing in $\mathbf{R}^n$. *Proc. A.M.S.* **97** (1986), 465–473.

[Ba3] ———. Logarithmically concave functions and sections of convex sets in $\mathbf{R}^n$. *Studia Math* **88** (1988), 69–84.

[Ba4] ———. Volumes of sections of cubes and related problems. GAFA (Israel Functional Analysis Seminar 87/88). Springer Lecture Notes in Math. 1376 (1989) 251–260.

[Ba5] ———. Normed spaces with a weak Gordon–Lewis property. Lecture Notes in Math. 1470 (1991) 36–47.

[Ba6] ———. Some remarks on the geometry of convex sets. (Israel Functional Analysis Seminar 86/87.) Springer Lecture Notes 1317 (1988), 224–231.

[B1] S. Bellenot. Tsirelson superspaces and ℓ_p. *Journal of Funct. Anal.* **69** (1986), 207–228.

[B2] ———. The Banach space T and the fast growing hierarchy from logic. *Israel J. Math.* **47** (1984), 305–313.

[Be] M. Berger. *Geometry*. Vol. II. Springer Verlag, 1986.

[BL] J. Bergh and J. Löfström. *Interpolation Spaces. An Introduction*. Springer Verlag (1976).

[Bl] W. Blaschke. Über affine Geometrie VII. Neue Extremeigenschaften von Ellipse und Ellipsoid. *Sitz. Ber. Akad. Wiss. Leipz. Math. Nat. Kl.* **69** (1917), 306–318.

[BF] T. Bonnesen and W. Fenchel. *Theorie der konvexen Körper*. Berlin, 1934.

[Bo]   C. Borell. The Brunn–Minkowski inequality in Gauss space. Inventiones Math. **30** (1975), 205–216.

[Bou1] J. Bourgain. On high dimensional maximal functions associated to convex bodies. *Amer. J. Math* **108** (1986), 1467–1476.

[Bou2] ———. On The L^p-bounds for maximal functions associated to convex bodies in $\mathbf{R}^n$. *Israel J. Math.* **54** (1986), 257–265.

[Bou3] ———. On finite dimensional homogeneous Banach spaces. GAFA (Israel Functional Analysis Seminar 86/87). Springer Lecture Notes 1317 (1988), 232–238.

[BCLT] J. Bourgain, P. Casazza, J. Lindenstrauss, and L. Tzafriri. *Banach Spaces with a Unique Unconditional Basis, Up to Permutation.* Memoirs of the A.M.S. n° 322 (1985), 1–111.

[BLM]   J. Bourgain, J. Lindenstrauss, and V. Milman. Approximation of zonoids by zonotopes. *Acta Math* **162** (1989), 73–141.

BMMP]   J. Bourgain, M. Meyer, V. Milman, and A. Pajor. On a geometric inequality. GAFA (Israel Functional Analysis Seminar 86/87). Springer Lecture Notes 1317 (1988), 224–231.

[BM]   J. Bourgain and V. D. Milman. New volume ratio properties for convex symmetric bodies in $\mathbf{R}^n$. *Inventiones Math.* **88** (1987), 319–340. See also: On Mahler's conjecture on the volume of a convex symmetric body and its polar preprint I.H.E.S., March 1985, and Sections euclidiennes et volume des corps convexes symétriques. *C. R. Acad. Sci. Paris.* A **300** (1985), 435–438.

[BoS]   J. Bourgain and S. Szarek. The Banach–Mazur distance to the cube and the Dvoretzky–Rogers factorization. *Israel J. Math.* **62** (1988), 169–180.

[BoT]   J. Bourgain and L. Tzafriri. Integrability of "large" submatrices with applications to the geometry of Banach spaces and harmonic analysis. *Israel J. Math.* **57** (1987), 137–224.

[BrL]   H. Brascamp, and E. Lieb. On extensions of the Brunn–Minkowski and Prékopa–Leindler including inequalities for log concave functions and with an application to the diffusion equation. *J. Funct. Anal.* **22** (1976), 366–389.

[BS]   A. Brunel and L. Sucheston. On J. Convexity and some ergodic super properties of Banach spaces. *Trans. Amer. Math. Soc.* **204** (1975), 79–90.

[BZ]   Y. Burago and V. Zalgaller. *Geometric Inequalities.* "Nauka," Leningrad, 1980, 288 pp. (Russian). Also published in English translation, Springer Verlag, 1988.

[Bu] D. Burkholder. Boundary value problems and sharp inequalities for martingale transforms. *Ann. Probab.* **12** (1984), 647–702.

[Ca1] B. Carl. Entropy numbers, *s*-numbers, and eigenvalue problems. *J. Funct. Anal.* **41** (1981), 290–306.

[Ca2] ———. Inequalities of Bernstein–Jackson type and the degree of compactness of operators in Banach spaces. *Ann. Inst. Fourier* **35** 3 (1985), 79–118.

[Ca3] ———. Entropy numbers of diagonal operators with an application to eigenvalue problems. *Journal Approx. Theory* **32** (1981), 135–150.

[CP] B. Carl and A. Pajor. Gelfand numbers of operators with values in a Hilbert space. *Inventiones Math* **94** (1988), 479–504.

[CT] B. Carl and H. Triebel. Inequalities between eigenvalues, entropy numbers and related quantities of compact operators in Banach spaces. *Math. Ann.* **251** (1980), 129–133.

[C] P. Casazza. Tsirelson's space. Proc. Research Workshop on Banach Space Theory (Iowa City, July 1981) edited by Bor-Luh Lin, University of Iowa Press, 1982, 9–22.

[CJT] P. Casazza, W. B. Johnson, and L. Tzafriri. On Tsirelson's space. *Israel J. Math.* **47** (1984), 81–98.

[CO] P. Casazza and E. Odell. Tsirelson's space and minimal subspaces. Longhorn Notes. University of Texas Functional Analysis Seminar, 82/83.

[CS] P. Casazza and T. Shura. *Tsirelson's Space.* Lecture Notes 1363 (1989). Springer Verlag.

[Ch] G. Chakerian. Inequalities for the difference body of a convex body. *Proc. A.M.S.* **18** (1967), 879–884.

[D] S. Dilworth. The cotype constant and large Euclidean subspaces of normed spaces. Preprint.

[DS] S. Dilworth and S. Szarek. The cotype constant and almost Euclidean decomposition of finite dimensional normed spaces. *Israel J. Math.* **52** (1985), 82–96.

[Di] C. Diminnie. A new orthogonality relation for normed linear spaces. *Math. Nachr.* **114** (1983), 197–203.

[Du1] R. M. Dudley. The sizes of compact subsets of Hilbert space and continuity of Gaussian processes. *J. Funct. Anal.* **1** (1967), 290–330.

[Du2] ———. Sample functions of the Gaussian process. *J. Funct. Anal.* **1** (1967), 290–330. Annals of Prob. **1** (1973).

[Dv] A. Dvoretzky. Some results on convex bodies and Banach spaces. Proc. Internat. Symp. on Linear Spaces. Jerusalem (1961), 123–160.

[DvR] A. Dvoretzky and C. A. Rogers. Absolute and unconditional convergence in normed linear spaces. *Proc. Nat. Acad. Sci.* (U.S.A.) **36** (1950), 192–197.

[Eg] H. Eggleston. *Convexity.* Cambridge University Press, 1958.

[Eh] A. Ehrhard. Inégalités isopérimétriques et intégrales de Dirichlet Gaussiennes. *Annales E.N.S.* **17** (1984), 317–332.

[El] J. Elton. Sign-embeddings of ℓ_1^n. *Trans A.M.S.* **279** (1983), 113–124.

[E] P. Enflo. On Banach spaces which can be given an equivalent uniformly convex norm. *Israel J. Math.* **13** (1972), 281–288.

[Fe1] X. Fernique. Régularité des trajectoires des fonctions aléatoires Gaussiennes. Ecole d'Eté de St Flour IV. 1974. Springer Lecture Notes in Maths. n° 480 (1975), 1–96.

[Fe2] _____. Intégrabilité des vecteurs gaussiens. *C. R. Acad. Sci.* A **270** (1970), 1698–1699.

[F] T. Figiel. A short proof of Dvoretzky's Theorem. *Composito Math.* **33** (1976), 297–301.

[FJ1] T. Figiel and W. B. Johnson. Large subspaces of ℓ_∞^n and estimates of the Gordon–Lewis constants. *Israel J. Math.* **37** (1980), 92–112.

[FJ2] _____. A uniformly convex space which contains no ℓ_p. *Compositio Math.* **29** (1974), 179–190.

[FLM] T. Figiel, J. Lindenstrauss, and V. D. Milman. The dimension of almost spherical sections of convex bodies. *Acta Math.* **139** (1977), 53–94.

[FT] T. Figiel and N. Tomczak-Jaegermann. Projections onto Hilbertian subspaces of Banach spaces. *Israel J. Math.* **33** (1979), 155–171.

[GaG] D. J. H. Garling and Y. Gordon. Relations between some constants associated with finite dimensional Banach spaces. *Israel J. Math.* **9** (1971), 346–361.

[GG] A. Garnaev and E. Gluskin. On diameters of the Euclidean Sphere. *Dokl. A.N. U.S.S.R.* **277** (1984), 1048–1052.

[Gl1] E. Gluskin. Norms of random matrices and diameters of finite dimensional sets. *Mat. Sbornik* **120** (1983), 180–189.

[Gl2] _____. Probability in the geometry of Banach spaces. (Russian) Proceedings of the I.C.M. Berkeley, U.S.A., 1986. Vol. 2, p. 924–938.

[Go1] Y. Gordon. Some inequalities for Gaussian processes and applications. *Israel J. Math.* **50** (1985), 265–289.

[Go2] ———. On Milman's inequality and random subspaces which escape through a mesh in $\mathbf{R}^n$. GAFA. Israel Funct. Analysis Seminar (86/87). Springer Lecture Notes 1317 (1988), 84–106.

[GKS] Y. Gordon, H. König, and C. Schütt. Geometric and probabilistic estimates for entropy and approximation numbers of operators. *Journal of Approx. Th.* **49** (1987), 219–239.

[GL] Y. Gordon and D. Lewis. Absolutely summing operators and local unconditional structures. *Acta. Math.* **133** (1974), 27–48.

[GMR] Y. Gordon, M. Meyer, and S. Reisner. Zonoids with minimal volume-product. A new proof. *Proc. A.M.S* **104** (1988), 273–276.

[GR] Y. Gordon and S. Reisner. Some aspects of volume estimates to various parameters in Banach spaces. Proc. of Research Workshop on Banach Space Theory (Iowa City, July 1981), edited by Bor–Luh Lin. University of Iowa Press, 1982, 23–54.

[Gr] M. Gromov. A note on the volume of intersection of balls. GAFA. Israel Funct. Analysis Seminar. Springer Lecture Notes 1267 (1987), 1–4.

[GM1] M. Gromov and V. Milman. A topological application of the isoperimetric inequality. *Amer. J. Math.* **105** (1983), 843–854.

[GM2] ———. Brunn Theorem and a concentration of volume phenomenon for symmetric convex bodies. GAFA. Israel Functional Analysis Seminar 1983-4. Tel Aviv University.

[GM3] ———. Generalization of the spherical isoperimetric inequality to the uniformly convex Banach spaces. *Composito Math.* **62** (1987), 263–282.

[G1] A. Grothendieck. *Produits Tensoriels Topologiques et Espaces Nucléaires.* Memoirs AMS **16** (1955).

[G2] ———. La théorie de Fredholm. *Bull. Soc. Math. France* **84** (1956), 319–384.

[G3] ———. La théorie de Fredholm. *Séminaire Bourbaki.* Mars 1984, exp. n° 91. 53/54. Paris, 1953.

[GW] P. Gruber and J. Wills. *Convexity and Its Applications.* Birkhäuser, 1983.

[Ha1] H. Hadwiger. *Altes und Neues über konvexe Körper.* Birkhäuser. Basel und Stuttgart, 1955.

[Ha2] ———. *Vorlesungen über Inhalt, Oberfläche und Isoperimetrie.* Springer Verlag, 1957.

[H] S. Heinrich. Ultraproducts in Banach space theory. *Journal für die Reine und angew. Math.* **313** (1980), 72–104.

[He] D. Hensley. Slicing the cube in $\mathbf{R}^n$ and probability. *Proc. A.M.S.* **73** (1979), 95–100.

[J] R. C. James. Some self-dual properties of normed linear spaces. *Annals of Math. Studies* n° 69 (1972).

[Joh] F. John. Extremum problems with inequalities as subsidiary conditions. Courant Anniversary Volume. Interscience, New York, 1948, 187–204.

[Jo1] W. B. Johnson. Personal Communication.

[Jo2] _____. A reflexive space which is not sufficiently Euclidean. *Studia Math.* **60** (1976), 201–205.

[Jo3] _____. Banach spaces all of whose subspaces have the Approximation Property. Séminaire d'Analyse Fonctionnelle 79/80, Ecole Polytechnique. Palaiseau. Exp. n° 16. Cf. also, Special Topics of Applied Mathematics. Functional Analysis, Numerical Analysis and Optimization. Proceedings Bonn 1979, edited by J. Frehse, D. Pallaschke, and V. Trottenberg, North Holland, 1980, 15–26.

[Jo4] _____. Homogeneous Banach Spaces. GAFA. Israel Functional Analysis Seminar 87/88. Springer Lecture Notes 1317 (1988), 201–203.

[JKMR] W. B. Johnson, H. König, B. Maurey, and J. R. Retherford. Eigenvalues of p-summing of ℓ_p-type operators in Banach spaces. *Journal of Funct. Anal* **32** (1979), 353–380.

[Kah1] J. P. Kahane. Une inégalité du type de Slepian et Gordon sur les processus Gaussiens. *Israel J. Math.* **55** (1986), 109–110.

[Kah2] _____. *Some Random Series of Functions.* Second edition. Cambridge University Press, 1985.

[Ka1] B. Kašin. Sections of some finite dimensional sets and classes of smooth functions. *Izv. Acad. Nauk* SSSR **41** (1977), 334–351. (Russian)

[Ka2] _____. On a certain isomorphism on $L^2(0,1)$. (Russian) *Comptes Rendus Acad. Bulgare des Sciences* **38** (1985), 1613–1615.

[Ko] A. Kolmogorov. Asymptotic characteristics of some completely bounded metric spaces. *Dokl. Akad. Nauk. SSSR* **108**, n° 3 (1956), 585–589.

[KT] A. Kolmogorov and B. Tikhomirov. ε-entropy and ε-capacity of sets in functional spaces. *Uspekhi Mat. Nauk* **14**, Part 2 (86),

(1959), 3–86 (Amer. Math. Soc. Translations [2] **17** (1961), 277–364.)

[Kö] H. König. *Eigenvalue Distribution of Compact Operators.* Birkhäuser, 1986.

[KM] H. König and V. D. Milman. On the covering numbers of convex bodies. Israel Functional Analysis Seminar GAFA. Springer Lecture Notes **1267** (1987), 82–95.

[KRT] H. König, J. Retherford, and N. Tomczak–Jaegermann. On the eigenvalues of $(p,2)$—summing operators and constants associated to normed spaces. *Journal of Funct. Analysis* **37** (1980), 88–126.

[KPS] S. Krein, J. Petunin and E. Semenov. *Interpolation of Linear Operators.* Transl. Math. Monogr. **54**, Amer. Math. Soc., Providence, 1982.

[Kr] J. L. Krivine. Sur un théorème de Kašin. Séminaire d'Analyse Fonctionnelle 83/84, Université Paris 7.

[K1] S. Kwapień. Isomorphic characterizations of inner product spaces by orthogonal series with vector coeficients. *Studia Math.* **44** (1972), 583–595.

[K2] ———. On operators factorizable through L_p-space. *Bull. Soc. Math. France. Mém* **31–32**, (1972), 215–225.

[KP] S. Kwapień and A. Pelczyński. The main triangle projection in matrix spaces and its applications. *Studia Math.* **34** (1970), 43–68.

[LS] H. Landau and L. Shepp. On the supremum of a Gaussian process. *Sankhya* **A32** (1970), 369–378.

[Le] L. Leindler. On a certain converse of Hölder's inequality. II. *Acta Sci. Math.* **33** (1972), 217–223.

[L] D. Lewis. Ellipsoids defined by Banach ideal norms. *Mathematika* **26** (1979), 18–29.

[Li] V. Lidskii. Non-self adjoint operators with a trace. *Dokl. Akad. Nauk SSSR* **125** (1959), 485–487.

[LP] J. Lindenstrauss and A. Pelczyński. Absolutely summing operators in $\mathcal{L}_p$ spaces and their applications. *Studia Math.* **29** (1968), 275–326.

[LT1] J. Lindenstrauss and L. Tzafriri. *Classical Banach Spaces I. Sequence Spaces.* Springer Verlag, 1977.

[LT2] ———. *Classical Banach Spaces II.* Function spaces. Springer Verlag, 1979.

[Lo] G. Lozanovskii. Banach structures and bases. *Funct. Anal. and Its Appl.* **294** (1967).

[MPi] M. Marcus and G. Pisier. *Random Fourier series with Applications to Harmonic Analysis.* Annals. of Math. Studies n° 101. Princeton Univ. Press, 1981.

[Ma1] Maurey, B. Un théorème de prolongement. *C. R. Acad. Sci. Paris*, A**279** (1974), 329–332.

[MaP] B. Maurey and G. Pisier. Séries de variables aléatoires vectorielles indépendantes et propriétés géométriques des espaces de Banach. *Studia Math.* **58** (1976), 45–90.

[Me] M. Meyer. Une caractérisation volumique de certains espaces normés de dimension finie. *Israel J. Math.* **55** (1986), 317–326.

[MeP] M. Meyer and A. Pajor. Volume des sections de la boule unité de ℓ_n^p. *J. Funct. Anal.* **80** (1988) 109–123.

[M1] V. D. Milman. Almost Euclidean quotient spaces of subspaces of finite dimensional normed spaces. *Proc. Amer. Math. Soc.* **94** (1985), 445–449.

[M2] _____. Volume approach and iteration procedures in local theory of normed spaces. Banach spaces. Proceedings, Missouri 1984, edited by N. Kalton and E. Saab, Springer Lecture Notes n° 1166 (1985), 99–105.

[M3] _____. Geometrical inequalities and mixed volumes in the local theory of Banach spaces. Colloque Laurent Schwartz. Astérisque. Soc. Math. France **131** (1985), 373–400.

[M4] _____. Random subspaces of proportional dimension of finite dimensional normed spaces: Approach through the isoperimetric inequality. Proceedings, Missouri 1984, edited by N. Kalton and E. Saab. Springer Lecture Notes n° 1166 (1985), 106–115.

[M5] _____. Inégalité de Brunn–Minkowski inverse et applications à le théorie locale des espaces normés. *C. R. Acad. Sci. Paris.* **302** Sér 1. (1986), 25–28.

[M6] _____. New proof of the theorem of Dvoretzky on sections of convex bodies. *Funkcional. Anal. i Prilozen* **5** (1971), 28–37.

[M7] _____. The concentration phenomenon and linear structure of finite dimensional normed spaces. Proceedings of the I.C.M. Berkeley, U.S.A. 1986. Vol. 2, p. 961–974.

[M8] _____. Isomorphic symmetrizations and geometric inequalities. GAFA. (Israel Functional Analysis Seminar) 86/87. Springer Lecture Notes 1317 (1988), 107–131.

[M9] _____. Entropy point of view on some geometric inequalities. *Comptes Rendus Acad. Sci. Paris* **306** (1988), 611–615.

[MPa] V. Milman and A. Pajor. Isotropic position and centroid body of the unit ball of a n-dimensional normed space. GAFA. Israel Functional Analysis Seminar. Springer Lecture Notes n° 1376 (1980), 64–104.

[MP1] Milman, V. D. and G. Pisier. Banach spaces with a weak cotype 2 property. *Israel J. Math.* **54** (1986), 139–158.

[MP2] _____. Gaussian processes and mixed volumes. Ann. of Prob. **15** (1987), 292–304.

[MS] Milman, V. D. and G. Schechtman. *Asymptotic theory of finite dimensional normed spaces.* Springer Lecture Notes n° 1200 (1986).

[Mi] B. Mitiagin. Approximative dimension and bases in nuclear spaces. *Uspehi Mat. Nauk* **16** (1961), 63–132.

[MiP] B. Mityagin and A. Pelczyński. Nuclear operators and approximative dimension. Proc. of ICM. Moscow (1966), 366–372.

[N] Neveu, J. *Processus Aléatoires Gaussiens.* Presses de l'Université de Montréal, 1968.

[Pa1] Pajor, A. Quotient volumique et espaces de Banach de type 2 faible. *Israel J. Math.* **57** (1987), 101–106.

[Pa2] _____. *Sous-espaces ℓ_n^1 des Espaces de Banach.* Travaux en cours. Hermann. Paris, 1986.

[Pa3] _____. Volumes mixtes et sous-espaces ℓ_1^n des espaces de Banach. Colloque en l'honneur de Laurent Schwartz. Astérisque Soc. Math. France **131** (1985), 401–411.

[PT1] Pajor, A. and N. Tomczak-Jaegermann. Subspaces of small codimension of finite dimensional Banach spaces. *Proc. Amer. Math. Soc.* **97** (1986), 637–642.

[PT2] _____. Nombres de Gelfand et sections enclidiennes de grande dimension. Séminaire d'Analyse Fonctionnelle 84/85. Publications de l'Université Paris 7.

[PT3] _____. Remarques sur les nombres d'entropie d'un opérateur et de son transposé. *C. R. Acad. Sci. Paris* **301** (1985), 743–746.

[PT4] _____. Volume ratio and other s-numbers of operators related to local properties of Banach spaces. *Journal of Funct. Anal.* **87** (1989), 273–293.

[Pe] Pelczyński, A. Geometry of finite dimensional Banach spaces and operator ideals. *Notes in Banach Spaces*, edited by E. Lacey, University of Texas Press, 1981.

[Pi1] Pietsch, A. *Operator ideals*. VEB. Berlin (1979) and North Holland (1980).

[Pi2] ———. Weyl numbers and eigenvalues of operators in Banach spaces. *Math. Ann.* **247** (1980), 149–168.

[Pi3] ———. *Eigenvalues and s-numbers*. Cambridge University Press, 1987.

[P1] Pisier, G. Probabilistic methods in the geometry of Banach spaces. CIME. Varenna, 1985. Springer Lecture Notes, n° 1206 (1986), 167–241.

[P2] ———. *Factorization of linear operators and Geometry of Banach spaces*. CBMS Regional Conference Series n° 60. AMS, 1986, p. 1–154. (Second printing with corrections, 1987).

[P3] ———. Holomorphic semi-groups and the geometry of Banach spaces. *Annals of Math.* **115** (1982), 375–392.

[P4] ———. Martingales with values in uniformly convex spaces. *Israel J. Math.* **20** (1975), 326–350.

[P5] ———. Weak Hilbert spaces. *Proc. London Math. Soc.* **56** (1988), 547–579.

[P6] ———. Quotients of Banach spaces of cotype q. *Proc. A.M.S.* **85** (1982), 32–36.

[P7] ———. Un théorème sur les opérateurs linéaires entre espaces de Banach qui se factorisent par un espace de Hilbert. *Ann. E.N.S.* **13** (1980), 23–43.

[P8] ———. A new approach to several results of V. Milman. *Journal für die Reine und Angew. Math.* **393** (1989) 115–131.

[P9] ———. Sur les espaces de Banach K-convexes. Séminaire d'Analyse Fonctionnelle 79/80. Ecole Polytechnique. Palaiseau. Exp. n°11.

[Pr] Prékopa, A. On logarithmically concave measures and functions. *Acta Sci. Math.* **34** (1973), 335–343.

[RS] Reed, M. and B. Simon. *Methods of Mathematical Physics*. Vol. 4. *Analysis of Operators*. Academic Press, New York, 1980.

[Re1] S. Reisner. Random polytopes and the volume product of symmetric convex bodies. *Math. Scand.* **57** (1985) 386–392.

[Re2] ———. Minimal volume-product in Banach spaces with a

1-unconditional basis. *Journal London Math. Soc.* **36** (1987), 126–136.

[Re3] ———. Zonoids with minimal volume product. *Math. Zeit.* **192** (1986) 339–346.

[Ro] M. Rogalski. Sur le quotient volumique d'un espace de dimension finie. Séminaire d'initiation à l'analyse 80/81. Université Paris 6, Paris.

[RSh] Rogers, C. A. and C. Shephard. The difference body of a convex body. *Arch. Math.* **8** (1957), 220–233.

[Ru] Ruston, A. *Fredholm Theory in Banach Spaces.* Cambridge University Press, 1986.

[StR1] Saint-Raymond, J. Sur le volume des corps convexes symétriques. Séminaire Initiation à l'Analyse. 80/81. Exp. n° 11, Université P. et M. Curie, Paris.

[StR2] ———. Le volume des idéaux d'opérateurs classiques. *Studia Math.* **80** (1984), 63–75.

[Sa1] Santaló, L. Un invariante afin para los cuerpos convexos del espacio de n dimensiones. *Portugal Math.* **8** (1949), 155–161.

[Sa2] ———. *Integral Geometry and Geometric Probability.* Encyclopedia of Maths., Vol. 1. Addison–Wesley, Reading, Mass., 1976. Reprinted Cambridge University Press, 1984.

[Sc1] C. Schütt. Entropy numbers of diagonal operators between symmetric Banach spaces. *Journal Approx. Theory* **40** (1984), 121–128.

[Sc2] ———. On the volume of unit balls in Banach spaces. *Compositio Math.* **47** (1982), 393–407.

[Si1] Simon, B. *Trace Ideals and Their Applications.* London Math. Soc. Lecture Notes n° 35. Cambridge University Press, 1979.

[Si2] ———. Notes on infinite determinants of Hilbert space operators. *Adv. Math.* **24** (1977), 244–273.

[Sl] Slepian, D. The one-sided barrier problem for Gaussian noise. *Bell System Tech. J.* **41** (1962), 463–501.

[Su] Sudakov, V. N. Gaussian processes and measures of solid angles in Hilbert space. *Soviet Math. Dokl.* **12** (1971), 412–415.

[S] Szarek, S. On Kašin's almost Euclidean orthogonal decomposition of ℓ_1^n. *Bull. Acad. Polon. Sci.* **26** (1978), 691–694.

[ST] Szarek, S. and N. Tomczak-Jaegermann, On nearly Euclidean decompositions of some classes of Banach spaces. *Compositio Math.* **40** (1980), 367–385.

[Sz]   A. Szankowski. On Dvoretzky's Theorem on almost spherical
       sections of convex bodies. *Israel J. Math.* **17** (1974), 325–338.

[Ta]   M. Talagrand. Regularity of Gaussian processes. *Acta Math.*
       159 (1987), 99–149.

[TJ1] N. Tomczak–Jaegermann. *Banach–Mazur Distances and Fi-
       nite Dimensional Operator Ideals.* Pitman, 1988.

[TJ2] ———. Dualité des nombres d'entropie pour des opérateurs
       à valeurs daus un espace de Hilbert. *C. R. Acad. Sci. Paris*
       t.305, Serie I (1987), 299–301.

[TJ3] ———. Computing 2-summing norms with few vectors. *Ark.
       Mat.* **17** (1979), 173–177.

[To]   Y. Tong. *Probability inequalities in multivariate distributions.*
       Academic Press. New York. 1980.

[T]   B. Tsirelson. Not every Banach space contains ℓ_p or c_o. *Funct.
       Anal. and Its Appl.* **8** (1974), 138–141 (translated from Rus-
       sian).

[Tz]   L. Tzafriri. On the type and cotype of Banach spaces. *Israel
       J. Math.* **32** (1979), 32–38.

[U]   Urysohn, P. S. Mean width and volume of convex bodies in an
       n dimensional space. *Mat. Sb.* **31** (1924), 477–486.

[V]   J. Vaaler. A geometric inequality with applications to linear
       forms. *Pacific J. Math.* **83** (1979), 543–553.

Index

Printed in the United States
By Bookmasters